Ben Stacy Jerrik (Ed.)

Rolland

Ben Stacy Jerrik (Ed.)

Rolland

San Sebastián Grand Prix

Part Press

Publisher:
Part Press is a trademark of
International Book Market Service Ltd., 17 Rue Meldrum, Beau Bassin, 1713-01 Mauritius
Email: info@bookmarketservice.com
Website: www.bookmarketservice.com

Published in 2012

Printed in: U.S.A., U.K., Germany. This book was not produced in Mauritius.

ISBN: 978-613-8-84222-4

Contents

Articles

Rolland-Pilain 1

San_Sebastián_Grand_Prix 2

24_Hours_of_Le_Mans 3

Albert_Guyot 16

Grand_Prix_motor_racing 17

World_War_I 24

Hydraulic_brake 77

Tours 82

References

Article Sources and Contributors 92

Image Sources, Licenses and Contributors 95

Rolland-Pilain

Industry	Automotive
Fate	ceased production
Founded	1905
Defunct	1932
Headquarters	Tours, France
Key people	François Rolland and Emile Pilain, founders
Products	Automobiles

Rolland-Pilain was a French car maker founded 1905 in Tours by François Rolland and Emile Pilain.[1]

The Rolland-Pilain cars pioneered many innovative solutions despite very limited financial resources. One such feature was hydraulic brakes. The models ranged from family cars to sports and luxury cars.

During World War I Rolland-Pilain produced shells and parts for aircraft engines using a female work force.

Participation in motor racing included several Grand Prix races and Le Mans 24 hours. The greatest success came in 1923, when Albert Guyot drove a Rolland-Pilain A22 to victory in the San Sebastián Grand Prix.[2]

A lack of funds led to Rolland and Pilain losing control of the company in 1926. Things went downhill from there and the factory closed in 1932. When production ceased, more than 5 000 cars had been built.

References

[1] Les Amis de ROLLAND-PILAIN (http://www.rolland-pilain.com)
[2] I Gran Premio de San Sebastián, Racing Database (http://www.racing-database.com/Race1.asp?GP=I Gran Premio de San Sebastián)

San_Sebastián_Grand_Prix

The **San Sebastián Grand Prix** (*Gran Premio de San Sebastián*) was an automobile race held at the Circuito Lasarte near San Sebastián, Spain.

During most of its existence, it was the main race event in Spain, since the Spanish Grand Prix had not yet been properly established as a grand prix race. The San Sebastián Grand Prix is sometimes retroactively counted as the Spanish Grand Prix, for those years when there was no Spanish Grand Prix held.

The eighth Grand Prix of San Sebastian was scheduled for July 25, 1930 but had to be cancelled due to the bad economic situation following the Wall Street crash in October 1929. Efforts were made and AIACR granted permission to still go ahead with the race on October 5, 1930. This was to be the last San Sebastián Grand Prix. When racing returned to the Lasarte circuit in 1933 it was as the Spanish Grand Prix.

Winners of the San Sebastián Grand Prix

These are the winners of all San Sebastián Grands Prix.[1] For other races held at the Lasarte circuit, see the Spanish Grand Prix.

Year	Driver	Constructor	Location	Report
1930	Achille Varzi	Maserati	Lasarte	Report
1929	Louis Chiron	Bugatti	Lasarte	Report
1928	Louis Chiron	Bugatti	Lasarte	Report
1927	Emilio Materassi	Bugatti	Lasarte	Report
1926	Jules Goux	Bugatti	Lasarte	Report *
1925	Albert Divo André Morel	Delage	Lasarte	Report
1924	Henry Segrave	Sunbeam	Lasarte	Report
1923	Albert Guyot	Rolland-Pilain	Lasarte	Report

* The 1926 race had the status of European Grand Prix

References

[1] Grand Prix winners 1919-1933, The golden era of Grand Prix racing (http://www.kolumbus.fi/leif.snellman/gpw2.htm)

24_Hours_of_Le_Mans

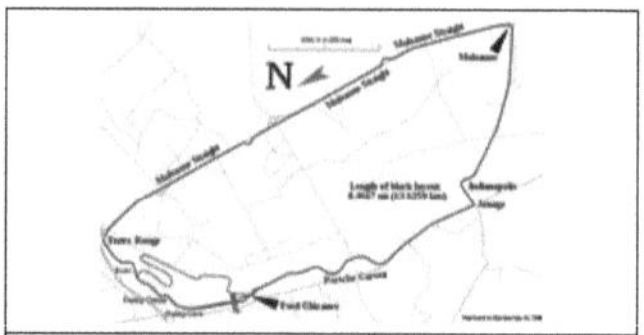

Venue	Circuit de la Sarthe
First race	1923
Duration	24 Hours
Most wins (driver)	Tom Kristensen (8)
Most wins (team)	Joest Racing (12)
Most wins (manufacturer)	Porsche (16)

The pits at dawn

The **24 Hours of Le Mans** (French: ***24 Heures du Mans***) is the world's oldest sports car race in endurance racing,[1] held annually since 1923 near the town of Le Mans, France. Commonly known as the Grand Prix of Endurance and Efficiency, race teams have to balance speed against the cars' ability to run for 24 hours without sustaining mechanical damage to the car and manage the cars' consumables, primarily fuel, tyres and braking materials. The endurance of the drivers is likewise tested as drivers frequently spend stints of over two hours behind the wheel before stopping in the pits and allowing a relief driver to take over the driving duties. Drivers then grab what food and rest they can before returning to drive another stint. Today it is mandated that three drivers share each competing vehicle.

The race is organised by the Automobile Club de l'Ouest (ACO) and runs on the Circuit de la Sarthe, a circuit containing a mix of closed public roads and specialist motor racing circuit that are meant not only to test a car and driver's ability to be quick, but also to last over a 24 hour period. The competing teams will race in groups called classes for cars of similar specification while at the same time competing for outright placing amongst all of the classes. Originally the race was held for cars as they were sold to the general public which were then called Sports Cars compared to the specialist racing cars used in Grands Prix. Over time the competing vehicles evolved away from their publicly-available road car roots and today the race is made of two classes specialised enclosed-bodywork two-seat Prototype sports cars and two classes of Grand Touring cars which bear much closer resemblance to high performance sports cars as sold to the public.[2]

Competing teams have had a wide variety of organisation, ranging from competition departments of road car manufacturers who are eager to prove the supremacy of their products, to professional motor racing teams who represent their commercial backers, some of which are also road car manufacturers attempting to win without the expense of setting up their own teams, to amateur race teams, racing as much to compete in the famous race as to claim victory for their commercial partners.

The race is held near the height of the European summer in June, leading at times to very hot weather conditions for the drivers, particularly in closed roof vehicles whose cabins can heat up to uncomfortably hot temperatures with generally poor ventilation; rain, however, is not uncommon. The race begins in mid-afternoon, racing through the night and following morning before finishing at the same time the race started, the following day.[3] Over the 24 hour period modern competitors will complete race distances well over 5000 km (**unknown operator: u'strong'** mi). The present record is 5410 km (**unknown operator: u'strong'** mi), recorded in the 2010 race.[4] It is a distance over six times longer than the Indianapolis 500, or approximately 18 times longer than a Formula One Grand Prix.

The race has over the years inspired imitating races all over the globe, popularising the 24 Hour format at places like Daytona, Nurburgring, Spa-Francorchamps, Sebring and Mount Panorama. Presently the American Le Mans Series and the European based Le Mans Series of multi-event sports car championships have been spun off from 24 Hours of Le Mans regulations. Other races include the Le Mans Classic, a race for historic Le Mans race cars of years past held on the Circuit de la Sarthe, a motorcycle version of the race which is held on the shortened Bugatti version of the same circuit, a kart race (24 Heures Karting) and a truck race (24 Heures Camions).

The race has also spent long periods as a round of the World Sportscar Championship, although Le Mans has always had a stronger reputation than the World Championship, and is presently a round of the FIA World Endurance Championship. The race is also known as a leg of the informal Triple Crown of Motorsport which links Formula One, IndyCars and Sports Car racing to represent a career achievement for drivers. Additionally it is seen as a leg of Triple Crown of endurance racing, which links the three largest sports car races together with 12 Hours of Sebring and 24 Hours of Daytona forming the other legs.

Purpose

At a time when Grand Prix racing was the dominant form of motorsport throughout Europe, Le Mans was designed to present a different test. Instead of focusing on the ability of a car company to build the fastest machines, the 24 Hours of Le Mans would instead concentrate on the ability of manufacturers to build sporty yet reliable cars. This encouraged innovation in producing reliable and fuel-efficient vehicles, because the nature of endurance racing requires cars that last the distance and spend as little time in the pits as possible.

At the same time, due to the layout of the Le Mans track, a need was created for cars to have better aerodynamics and stability at high speeds. While this was shared with Grand Prix racing, few tracks in Europe had straights of a length comparable to the Mulsanne. The fact that the road is public and therefore not maintained to the same quality as some permanent racing circuits also put more of a strain on parts, putting greater emphasis on reliability.

The demand for fuel economy created by the oil crisis in the early 1970s led the race organisers to adopt a fuel economy formula known as Group C, in which the amount of fuel each car was allowed to use during the race was limited. Although Group C was abandoned when teams were able to master the fuel formulae, fuel economy was still important to some teams as alternative fuel sources appeared in the early 21st century, attempting to overcome time spent during pit stops.

These technological innovations have had a trickle-down effect, with technology used at Le Mans finding its way into production cars several years later. This has also led to faster and more exotic supercars due to manufacturers wishing to develop faster road cars for the purposes of developing them into even faster GT cars.

The race

Cars

The total entry has usually consisted of approximately 50 competitors. Each car is required to have at least two seats, although in recent years only the ability to place a second seat in the cockpit has been required; the seat itself has not. No more than two doors are allowed; open cockpit cars do not require doors.

Although all cars compete at the same time, there are separate classes. A prize is awarded to the winner of each class, and to the overall winner.

An Aston Martin DBR9, entrant in the former LMGT1 class

The number of classes has varied over the years, but currently there are four. Custom-built Le Mans Prototypes (LMP) are the top two classes, LMP1 and LMP2, divided by speed, weight, and power output. From 2011, the next two classes are production-based grand tourer (GT) classes, GT Endurance Pro and GT Endurance AM. Both of these classes utilise the FIA GT2 class. Although the top class is the most likely to provide the winner of the race, lower classes have won on occasion due to better reliability.

Drivers

Originally, there were no rules on the number of drivers of a car, or how long they could drive. Although almost all teams used two drivers in the early decades, some Le Mans drivers such as Pierre Levegh and Eddie Hall attempted to run the race solo, hoping to save time by not having to change drivers. This practice was later banned. Until the 1980s, there were teams in which only two drivers competed, but by the end of the decade the rules were changed to stipulate that at least three drivers must drive each car.

By the 1990s, due to the speeds of the cars and the strain it put on drivers, further rules were added to improve driver safety. Drivers could not drive more than four hours consecutively, and no one driver could run for more than fourteen hours in total. This reduced driver fatigue during the races.

Unique rules and traditions

Although the 24 Hours of Le Mans was part of the World Sportscar Championship for most of its existence, it has regularly had rules which differed from those used in other series, partly due to the length of the event. Some rules are for safety reasons, while others are for the purposes of competition.

For many decades, cars were required to run at least an hour into the race before they were allowed to refill fluids for the car, such as oil or coolant, with the exception of fuel. This was an attempt by the ACO to help increase efficiency and reliability. Cars which could not last the first hour without having to replace lost fluids were disqualified.

Another rule that is unique to Le Mans is a requirement for cars to be shut off while they are being refueled in the pits. Based not only on the notion that it is safer and less of a fire hazard to do so, this also allows for another test of reliability, because cars have to test their ability to restart many times under race conditions. Another element of this rule is that mechanics are not allowed to work on the car or its tyres while it is being refueled, which has led teams to adapt innovative ways in which to decrease the time of these lengthy pit stops. As an exception to this rule, drivers are allowed to get out of the car and be replaced by another driver during refueling.

At Le Mans there are various traditions. One of the longest lasting is the waving of the French tricolor to start the race. This is usually followed by a fly-over featuring jets trailing blue, white and red smoke. A similar flag tradition is the waving of safety flags during the final lap of the race by track marshals, congratulating the winners and other finishers.

The 24 Hours of Le Mans was the venue for the first known instance at a major race of a winning driver celebrating by spraying champagne instead of drinking it.[5] When Dan Gurney won the 1967 race with co-driver A.J. Foyt, the two drivers mounted the victory stand and Gurney was handed a magnum of champagne. Looking down, he saw Ford CEO Henry Ford II, team owner Carroll Shelby and their wives, as well as several journalists who had predicted disaster for the high-profile duo. Gurney shook the bottle and sprayed everyone nearby, establishing a tradition re-enacted in victory celebrations the world over for the next 40+ years. Gurney, incidentally, autographed and gave the bottle of champagne to a LIFE magazine photographer, Flip Schulke, who used it as a lamp for many years. He recently returned the bottle to Gurney, who keeps it at his home in California.

Schedule

The first race was held on 26 and 27 May 1923 and has since been run annually in June with exceptions occurring in 1956, when the race was held in July and 1968, when it was held in September due to nationwide political turmoil earlier that year (see May 1968). The race has been cancelled twice: once in 1936 (labour strike during the Great Depression) and from 1940 to 1948 (World War II and its aftermath).

The race weekend also usually takes place on the second weekend of June, with qualifying and practice taking place on the Wednesday and Thursday before the race, following an administrative scrutinizing of the cars on Monday and Tuesday. Currently these sessions are held in the evening, with two separate two hour sessions held each night. A day of rest is scheduled on Friday, and includes a parade of all the drivers through the centre of the town of Le Mans.

A test day was also usually held prior to the event, traditionally at the end of April or beginning of May. These test days served as a pre-qualification for the event, with the slowest cars not being allowed to appear again at the proper qualifying. However, with the cost necessary to transport cars to Le Mans and then back to their respective series in between the test and race weeks, the test day was moved to the first weekend of June for 2005. The notion of pre-qualifying was also eliminated in 2000, when all competitors invited to the test would be allowed into the race.

The Le Mans Legend races have also been part of the schedule since 2001, usually running exhibition races during qualifying days, a few hours prior to the sessions for the Le Mans entrants.

Traditionally until 2008, the race started at 16:00 on the Saturday, although in 1968 the race started at 14:00 due to the lateness of the race on the calendar. In both 1984 and 2007, the start time was moved ahead to 15:00 due to the conflicting French General Election. In 2006, the ACO scheduled a 17:00 start time on Saturday, 17 June in order to maximize television coverage in between the FIFA World Cup games. In 2009 the race took place over 13–14 June and started at 15:00 local time (13:00 GMT), and has started at that time ever since.

Classification

Rolling start of the 2008 race

Originally, the race results were determined by distance. The car which covered the greatest distance was declared the winner. This is known to have caught out the Ford team in 1966. With a dominant 1–2 lead, the two cars slowed to allow for a photo opportunity at the finish line, with Denny Hulme slightly ahead of Bruce McLaren. However, since McLaren's car had actually started much farther back on the grid than Hulme, McLaren's car had actually covered the greatest distance over the 24 hours. With the margin of victory determined to be eight metres, McLaren and his co-driver, Chris Amon, were declared the winners. Not only did the decision cost Hulme a victory it also deprived his co-driver Ken Miles of the win. Miles had all ready won the other two endurance races at Sebring and Daytona. With a win at Le Mans he would have become the first man to win all three not to mention in the same year. Miles was one of the oldest racer on the circuit. Tragically he was killed in a crash later that year. The greatest distance rule was later changed when a rolling start was introduced, and instead the winner is now the car that has completed the greatest number of laps.

To be classified in the race results, a car is required to cross the finish line after 24 hours. This has led to dramatic scenes where damaged cars wait in the pits or on the edge of the track close to the finish line for hours, then restart their engines and crawl across the line to be listed amongst the finishers. However, this practice of waiting in the pits was banned in recent years with a requirement that a team complete a set distance within the last hour to be classified as a finisher.

Another rule instituted by the ACO was the requirement that cars complete 70% of the distance covered by the overall winner. A car failing to complete this number of laps, even if it finished the race, was not deemed worthy of classification because of poor reliability or speed.

Le Mans start

Further information: standing start

The race traditionally began with what became known as the *Le Mans start*, in which cars were lined up alongside the pit wall in the order in which they qualified. The starting drivers would stand on the opposite side of the front stretch. When the French flag dropped to signify the start, the drivers would run across the track to their cars, which they would have to enter and start without assistance, before driving away. This became a safety issue in the late 1960s when drivers would ignore their safety harnesses, which were then a recent invention. This led to drivers running the first few laps either improperly harnessed due to attempting to do it while driving or sometimes not even harnessed at all, leading to several deaths when cars were involved in accidents due to the bunched field at the start.

The permanent pits and pit straight for both the *Circuit de la Sarthe* and Bugatti Circuit

This starting method inspired Porsche to locate the ignition key switch to the left of the steering wheel. In a left-hand drive car, this allowed the driver to use his left hand to start the engine, and his right hand to put the transmission into gear, which in turn shaves off a few tenths of a second.

Another method for speeding up the start was developed by Stirling Moss. His car was waiting with 1st gear already engaged. When he jumped in, he switched the starter on without depressing the clutch. The car was immediately jerked forward by the starter motor, but the engine did not start due to low RPM. After a few seconds of motion, he then pushed the clutch down, allowing the engine to speed up and start while the car was moving.

Feeling this type of start was unsafe, in the 1969 race Jacky Ickx opposed it by walking across the track while his competitors ran. Although he was nearly hit by a faster competitor's car while walking, Ickx took the time to fasten his safety belts before pulling away. Privateer John Woolfe was killed in an accident on the first lap of that race. Ickx went on to win.

The traditional Le Mans start was changed for 1970. Cars were still lined up along the pit wall, but the drivers were already inside and strapped in. At the dropping of the French tricolor, the drivers would then start their engines and drive away. However, in 1971 this method was done away with altogether, and instead a rolling start (sometimes known as an *Indianapolis start*) was introduced and has been used ever since.

The circuit

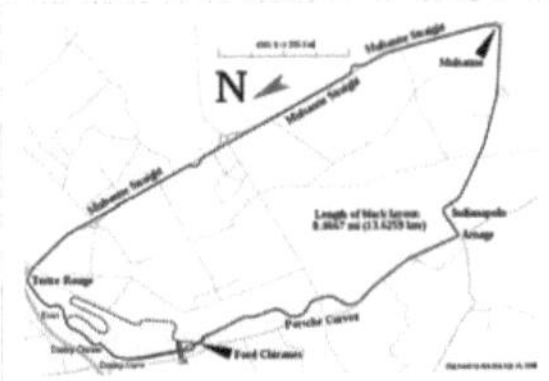

The Circuit de la Sarthe with the Bugatti Circuit in gray

The circuit on which the 24 Hours of Le Mans is run is named the Circuit de la Sarthe (Circuit of the Sarthe), after the department that Le Mans is within. It consists of both permanent track and public roads that are temporarily closed for the race. Since 1923 the track has been extensively modified, mostly for safety reasons, and currently is 13.629 km in length. Although it initially entered the town of Le Mans, the track was cut short in order to better protect spectators. This led to the creation of the Dunlop Curve and Tertre Rouge corners before rejoining the old circuit on the Mulsanne. Another major change was on the Mulsanne itself, when the FIA decreed that it would no longer sanction any circuit that had a straight longer than 2 km. This led to the addition of two chicanes, reducing the time that the cars spent travelling at very high speeds on the old 6 km long straight.

The public sections of the track differ from the permanent circuit, especially in comparison with the Bugatti Circuit which is inside the Circuit de la Sarthe. Due to heavy traffic in the area, the public roads are not as smooth or well kept. They also offer less grip because of the lack of soft tyre rubber laid down from racing cars, though this only affects the first few laps of the race. The roads are closed only within a few hours of the practice sessions and the race, before being opened again almost as soon as the race is finished. Workers have to assemble and dismantle safety barriers every year for the public sections.

History

For a list of individual race reports, see Category:24 Hours of Le Mans races.

1923–1939

A poster for the 1923 24 Hours of Le Mans

The 24 Hours of Le Mans was first run on 26 and 27 May 1923, through public roads around Le Mans. Originally planned to be a three year event awarded the Rudge-Whitworth Triennial Cup, with a winner being declared by the car which could go the farthest distance over three consecutive 24 Hour races, this idea was abandoned in 1928 and overall winners were declared for each single year depending on who covered the farthest distance by the time 24 hours were up. The early races were dominated by French, British, and Italian drivers, teams, and cars, with Bugatti, Bentley, and Alfa Romeo being the dominant marques. Innovations in car design began appearing at the track in the late 1930s, with Bugatti and Alfa Romeo running highly aerodynamic bodywork in order to run down the Mulsannes Straight at faster speeds. In 1936 the race was cancelled due to general strikes in France, then with the outbreak of World War II in late 1939, the race went on a ten-year hiatus.

1949–1969

Following the reconstruction of the circuit facilities, the race was resumed in 1949 with renewed interest from major automobile manufacturers. After the formation of the World Sportscar Championship in 1953, of which Le Mans was a part, Ferrari, Aston Martin, Mercedes-Benz, Jaguar, and many others began sending multiple cars backed by their respective factories to compete for overall wins against their competitors. Their competition sometimes resulted in tragedy, as in an accident during the 1955 race in which Pierre Levegh's car crashed into a crowd of spectators, killing more than 80 people. The incident led to the widespread introduction of safety measures, not only at the circuit but elsewhere in the motorsports world. However, even though the safety standards improved, so did the speed of the cars; the move from open-cockpit roadsters to closed-cockpit coupes resulted in speeds of over 320 kilometres per hour (**unknown operator: u'strong'** mph) on the Mulsanne. Race cars of the time were still mostly based on production road cars, but by the end of the 1960s, Ford had entered the picture with their GT40s, taking four straight wins before the era of production-based wins came to a close.

1970–1980

Renault Alpine A443 from 1978

For the new decade, the race took a turn towards more extreme speeds and automotive designs. These extreme speeds led to the replacement of the typical standing *Le Mans start* with a rolling *Indianapolis start.* Although production-based cars still raced, they were now in the lower classes while purpose-built sportscars became the norm. The Porsche 917, 935, and 936 were dominant throughout the decade, but a resurgence by French manufacturers Matra-Simca and Renault saw the first victories for the nation since the 1950 race. This decade is also remembered for strong performances from many privateer constructors, with two scoring the only victories for a privateer. John Wyer's Mirage won in 1975 while Jean Rondeau's self-titled chassis took 1980.

1981–1993

The dominant Group C formula Porsche 962s

The rest of the 1980s was known for the dominance by Porsche under the new Group C race car formula that encouraged fuel efficiency. Originally running the effective 956, it was later replaced by the 962. Both chassis were affordable enough for privateers to purchase them en masse, leading to the two model types winning six years in a row. Jaguar and Mercedes-Benz returned to sports car racing, with Jaguar being the first to break Porsche's dominance with victories in 1988 and 1990 (with the XJR-9 and Jaguar XJR-12 respectively). Mercedes-Benz won in 1989, with what was seen as the latest incarnation of the elegant "Silver Arrows", the Sauber C9, while an influx of Japanese manufacturer interest saw prototypes from Nissan and Toyota. In 1989 too a W.M.-Peugeot set up a new record speeding at 406 km/h (253 mph) in the *Ligne Droite des Hunaudières*, famous for its 6 km (3.7 mi) long straight. Mazda would be the only Japanese manufacturer to succeed, with their unique rotary-powered 787B winning in 1991. For 1992 and 1993, Peugeot entered the sport and dominated the race with the Peugeot 905 as the Group C formula and World Sportscar Championship were fading in participation.

The circuit would also undergo one of its most notable changes in 1990, when the 5 km long Mulsanne was modified to include two chicanes in order to stop speeds of more than 400 km/h (**unknown operator: u'strong'** mph) from being reached. This began a trend by the ACO to attempt to slow the cars on various portions of the track; although speeds over 320 km/h (**unknown operator: u'strong'** mph) are still regularly reached at various points on a lap.

1994–1999

Following the demise of the World Sportscar Championship, Le Mans saw a resurgence of production-based grand tourer cars. Thanks to a loophole in the rules, Porsche succeeded in convincing the ACO that a Dauer 962 Le Mans supercar was a production car, allowing Porsche to race their Porsche 962 for one final time, dominating the field. Although the ACO attempted to close the loop hole for 1995, newcomer McLaren would win the race in their supercar's first appearance thanks to reliability, beating faster yet more trouble prone prototypes. The trend would continue through the 1990s as more exotic supercars were built in order to skirt the ACO's rules regarding production-based race cars, leading to Porsche, Mercedes-Benz, Toyota, Nissan, Panoz, and Lotus entering the GT categories. This culminated in the 1999 event, in which these GT cars were faced with the Le Mans Prototypes of BMW, Audi, and Ferrari. BMW would survive with the victory, their first ever.

This strong manufacturer influence led the ACO to lending the Le Mans name to a sports car series in the United States in 1999, known as the American Le Mans Series, which competes to this day and serves to qualify teams to enter Le Mans.

2000–2005

A diesel-powered Audi R10 TDI

Many major automobile manufacturers withdrew from sports car racing after the 1999 event, because of the cost involved. Only Cadillac and Audi remained, and Audi easily dominated the race with their R8. Cadillac pulled out of the series after three years, and although Panoz, Chrysler, and MG all briefly attempted to take on Audi, none could match the R8's performance. After three victories in a row, Audi provided engine, team staff and drivers to their corporate partner Bentley, who had returned in 2001, and the factory Bentley Speed 8s were able to succeed ahead of privateer Audis in 2003.

Since 2006

At the end of 2005, after five overall victories for the R8, and six to its V8 turbo engine, Audi took on a new challenge by introducing a diesel engined prototype known as the R10 TDI. Although not the first diesel to race, it was the first to win at Le Mans. This era saw other alternative fuel sources being used, including bio-ethanol, while Peugeot decided to follow Audi's lead and also pursue a diesel entry in 2007 with their 908 HDi FAP.

The 2008 24 Hours of Le Mans was a great race between the Audi R10 TDI and the Peugeot 908 HDi FAP. After 24 hours of racing, the Audi managed to win the race by a margin of less than 10 minutes. For the 2009 24 Hours of Le Mans, Peugeot introduced a new energy-recovery system similar to the KERS used in Formula One.[6] Aston Martin entered the LMP1 category, but still raced in GT1 with private teams. Audi returned with the new R15 TDI, but this time Peugeot prevailed, taking their first overall triumph since 1993.

A second ACO-backed series was also formed, similar to the American Le Mans Series, but concentrating on Europe. The Le Mans Endurance Series (later shortened to Le Mans Series), resurrected many well known 1000 kilometres (**unknown operator: u'strong'** mi) endurance races, and was followed by the Asian-centered Japan Le Mans Challenge in 2006.

In 2011, the race became the premier round of the Intercontinental Le Mans Cup, an attempt to make a world championship for Endurance Racing again. In 2012, the race became the centerpiece of the FIA World Endurance Championship, the successor to the ILMC.

As of 2011, Porsche remains the most successful manufacturer with a record 16 overall victories, including a record seven in a row.

Innovations

Over its lifetime, Le Mans has seen many innovations in automotive design in order to counteract some of the difficulties that the circuit and race present. These have either been dictated by rules or have been attempts by manufacturers to outwit the competition. Some have made their way into the common automobile and are used nearly every day.

Aerodynamics

A Porsche 908 *Langheck*, German for "Long Tail"

One of the keys to Le Mans is top speed, caused by the long straights that dominate the circuit. This has meant cars have attempted to achieve the maximum speeds possible instead of relying on downforce for the turns. While early competitors cars were street cars with their bodywork removed to reduce weight, innovators like Bugatti developed cars which saw the beginnings of aerodynamics. Nicknamed *tanks* due to their similarity to a drop tank, these cars used simple curves to cover all the mechanical elements of the car and increase top speed. Once Le Mans returned after World War II, most manufacturers would adopt closed bodies which were streamlined for better aerodynamics. A notable example in the changes brought about by aerodynamics are the 1950 entries by Briggs Cunningham. Cunningham entered two 1950 Cadillac Coupe de Villes, one nearly stock and the other completely rebodied in a streamlined aluminum shape developed by aeronautical engineers from Grumman Aircraft Engineering Corporation. The streamlined car looked so unusual that it was nicknamed "Le Monstre" by the French press. The smoothing of body shapes and fairing-in of various parts of the machine brought about by the continual search for reduction of aerodynamic drag led to a separation from Grand Prix cars, which rarely had large bodywork.

As the years went on, bodywork became larger while at the same time lighter. The larger bodywork was able to provide more downforce for the turns without increasing the drag, allowing cars to maintain the high top speeds. These extended bodyworks would usually concentrate on the rear of the car, usually being termed *long tail*. The bodywork also began to cover the cockpit for less drag, although open cockpits would come and go over the years as rules varied. Aerodynamics reached its peak in 1989, before the Mulsanne Straight was modified. During the 1988 race, the crew of a W.M. prototype taped over the engine openings and set a recorded speed of 404 km/h (251 mph) down the Mulsanne in a publicity stunt, although the car was almost undrivable elsewhere on the circuit and the engine was soon destroyed from a lack of cooling. However, for the 1989 event, the Mercedes-Benz C9 reached 399 km/h (248 mph) under qualifying conditions.

Engines

An early supercharged Bentley

A wide variety of engines have competed at Le Mans, in attempts to not only achieve greater speed but also to have better fuel economy, and spend less time in the pits. Engine sizes have also varied greatly, with the smallest engines being a mere 569 cc (Simca Cinq) and the largest upwards of 7986 cc (Chrysler Viper GTS-R). Supercharging was an early innovation for increasing output, first being raced in 1929, while turbocharging would not appear until 1974.

The first car to enter without an engine run by pistons would be in 1963, when Rover partnered with British Racing Motors to run a gas turbine with mixed success, repeating again in 1965. The American Howmet Corporation would attempt to run a turbine again in 1968 with even less success. Although the engines offered great power, they were notoriously hot and uneconomical for fuel.

Another non-piston engine that would appear would be a Wankel engine, otherwise known as the rotary engine. Run entirely by Mazda since its introduction in 1970, the compact engine would also suffer from fuel economy problems like the turbine had, yet would see the success that the turbine lacked. After many years of development, Mazda finally succeeded in being the only winner of the race to not have a piston-powered engine, taking the 1991 event with the 787B.

Alternative fuel sources would also play a part in more normal engine designs, with the first non-gasoline car appearing in 1949. The Delettrez Special would be powered by a diesel engine, while a second diesel would appear in the form of the M.A.P. the following year. Although diesel would appear at other times over the race existence, it would not be until 2006 when a major manufacturer, Audi, would invest in diesels and finally succeed, with the R10 TDI.

Ethanol fuel appeared in 1980 in a modified Porsche 911, taking a class win. Alternative biological fuel sources would return again in 2004 with Team Nasamax's DM139-Judd.[7] In 2008, the use of biofuels (10% ethanol for petrol engines and biodiesel respectively for diesel engines) were allowed. Audi was the first to use next generation 10% BTL biodiesel manufactured from biomass and developed by partner Shell.[8]

From 2009 onwards the Le Mans regulations new from the ACO[9] allow hybrid vehicles to be entered, with either KERS or TERS (Kinetic/Thermal Energy Recovery System) setups, however the only energy storage allowed will be electrical (i.e. batteries), seemingly ruling out any flywheel-based energy recovery systems. Cars equipped with KERS systems were allowed to race in 2009 with specific classification rules. From 2010 they will be able to compete for points and the championship.

Brakes

With increased speeds around the track, brakes become a key issue for teams attempting to safely bring their cars down to a slow enough speed to make turns such as Mulsanne Corner. Disc brakes were first seen on a car when the Jaguar C-Type raced at Le Mans in 1953. The Mercedes-Benz 300 SLR would introduce the concept of an air brake in 1955, using a large opening hood on the rear of the car.

In the 1980s, anti-lock braking systems would become standard on most Group C cars as a safety measure, ensuring that cars did not lose control while still moving at approximately 320 km/h. By the late 1990s, reinforced carbon-carbon brakes would be adapted for better stopping power and reliability.

Successful marques and drivers

For a list of winning drivers, teams, and cars, see List of 24 Hours of Le Mans winners.

The most successful participant of all time at Le Mans, Danish driver Tom Kristensen has eight wins.

Over the years, many manufacturers have managed to take the overall win, while even more have taken class wins. By far the most successful marque in the history of the race is Porsche, which has taken sixteen overall victories, including seven in a row from 1981 to 1987. Audi is next with ten, and Ferrari follows with nine, also including six in a row from 1960 to 1965. Recently the Audi marque has dominated the event, winning in ten of the thirteen years it has participated. Audi and Team Joest have had two hat-tricks, the first being in 2000, 2001, and 2002. Jaguar has seven wins, while Bentley, Alfa Romeo, and Ford all managed to win four races in a row, with Bentley recording two other victories in other years as well. The only Japanese marque to win the race so far has been Mazda, although nearly every major Japanese manufacturer has made attempts at the race. Mazda's 1991 victory is the only win by a rotary engine, one of Mazda's hallmarks.

Two drivers stand apart for their number of victories. Initially Jacky Ickx held the record at six, scoring victories between 1969 and 1982, earning him an honorary citizenship to the town of Le Mans. However Dane Tom Kristensen has beaten this record with eight wins between 1997 and 2008, including six in a row. Three-time winner Woolf Barnato (1928 to 1930) and American racing legend AJ Foyt (1967) are still the only drivers to have won every Le Mans they participated in.

Henri Pescarolo has won the race four times, and currently holds the record for the most Le Mans appearances at 33. Japan's Yojiro Terada, currently still active as a driver, holds the record for the most Le Mans starts without a win. Graham Hill is the only driver to win the so-called *Triple Crown of Motorsport* which is defined as winning the Indianapolis 500 (won by Hill in 1966), Monaco Grand Prix (1963, 1964, 1965, 1968, 1969) and the 24 Hours of Le Mans (1972)[10] [11]

Accidents

With the high speeds associated with Le Mans, the track has seen a number of accidents, some of which have been fatal to drivers and spectators. The worst moment in Le Mans history was during the 1955 race in which more than 80 spectators and driver Pierre Levegh were killed. In the shock following this disaster, many major and minor races were cancelled in 1955, such as the Grand Prix races in Germany and Switzerland (the latter as a reaction having banned motorsport round-track races throughout the entire country, the ban was only lifted in 2007[12]). This accident brought wide sweeping safety regulations to all motorsports series, for both driver and spectator protection. In 1986 Jo Gartner drove a Porsche 962C and crashed into the barriers on the Mulsanne straight, killing him instantly. His accident was the most recent fatality in the race itself, however there was the fatality of Sebastien Enjolras in 1997 during the practices.

In one of the most recognizable recent accidents, calamity would once again strike Mercedes-Benz, although without fatality. The Mercedes-Benz CLRs which competed in 1999 would suffer from aerodynamic instabilities that caused the cars to become airborne in the right conditions. After initially happening at the Le Mans test day, Mercedes claimed they had solved the problem, only to have it occur again at Warm Up hours before the race. Mark Webber was the unlucky driver to flip the car on both occasions. The final and most damaging accident occurred during the race itself when Peter Dumbreck's CLR became airborne and then proceeded to fly over the safety fencing, landing in the woods several metres away. No drivers were badly hurt in any of the three accidents, but Mercedes-Benz quickly withdrew their remaining entry and ended their entire sportscar programme.

In 2011, two horrific looking accidents would occur to two of the three factory Audis running in the LMP1 class. Near the end of the first hour, the #3 car driven by Allan McNish collided with one of the Ferrari GT class cars resulting in McNish's car violently smashing into the tyre wall and being thrown into the air at the Dunlop chicanes, resulting in pieces of bodywork flying over and nearly hitting many photographers on the other side of the barrier. In the eleventh hour of the race, another massive accident would occur this time to the #1 car driven by Mike Rockenfeller when he also appeared to have contact with another Ferrari GT car. On the run up to Indianapolis corner Rockenfeller's Audi was sent into the outside barrier at well over 170 miles per hour (**unknown operator: u'strong'** km/h). Only the main cockpit safety cell of the car remained along with major damage being done to the barriers that needed to be repaired before the race was resumed. Nobody in either of these accidents was injured, despite both chassis' being written off, thus showing how racing cars continue to advance in safety over the years.

Appearances in media

The 1964 event plays a critical part in the Academy Award winning *Un Homme et Une Femme*, in which the wife of the driver hero commits suicide when she mistakenly thinks that he has been killed in an accident during the race.

The 1969 event, known for its close finish, was documented in a short film entitled *La Ronde Infernale*. This was given a limited cinema release but is now available on DVD.

The race became the center of a major motion picture in 1971 when Steve McQueen released his simply titled *Le Mans*, starring McQueen as Michael Delaney, a driver in the 1970 event for the Gulf Porsche team. Likened to other motorsports films such as *Grand Prix* for Formula One racing and *Winning* for the Indianapolis 500, Le Mans is the best known film to center on sports car racing. It was filmed during the race using modified racing cars carrying cameras, as well as purchased Porsche 917s, Ferrari 512s and Lola T70s for action shots made after the race. The Porsche 908 which served as a camera car in the race actually finished, yet was so far behind the winners due to lengthy reel changes during pit stops that it was not classified in the results.

A modern film not centering on Le Mans yet featuring events from the 2002 race was *Michel Valliant*, about a French comic book motorsports hero. Again using two camera cars to tape action during the race, the French film was not as widely accepted as *Le Mans* had been. The 1974 TV show *The Goodies* also featured an episode entitled *The Race*, involving a comedic trio attempting to run Le Mans.

More recently, a documentary film called *Truth in 24*, narrated by Jason Statham, covered the Audi team in its effort to win a 5th straight title in 2008. The race features prominently as the film covers the racing season leading up to the Le Mans race.

The race has also been used for several video games over the years, some of which have allowed players to compete for the full 24 hours. Although most used the Le Mans name itself, the PlayStation 2 game *Gran Turismo 4* also included the Circuit de la Sarthe and allowed players to run the full 24 hour races with and without the chicanes on the Mulsanne Straight. The Xbox 360, PlayStation 3 and PC game *Race Driver: Grid* also includes the 24 Hours of Le Mans at the end of each in-game season albeit being only 24 minutes in length by default. However the player can also choose to compete in the race for different lengths of time ranging from several minutes to a full 24 hours.

Coverage

Motors TV covered the Le Mans 24 Hours in the entirety in 2006 and 2007. This included coverage of the scrutineering, qualifying, driver parade, warm up and the whole race. In the United States, Speed Channel airs partial live coverage through a combination of coverage from the French host broadcaster and their own pit reporting crew. In 2008 Eurosport secured a multi-year deal to show the entire race including the qualifying and the motorcycle race. Every hour of the 2008 race was broadcast in segments on the main channel and on Eurosport 2, however in recent years, a couple of Hours have been missed due to scheduling clashes with other sports.[13] In addition live streaming video was provided on Eurosport's web page, albeit not for free. But since 2009, Eurosport and Eurosport 2 has been covering non-stop between those two channels, all 24 hours of action.

The race is also broadcast (in English) on radio by Radio Le Mans. Broadcast from the circuit for the full 24 hours as well as before and after, it offers fans at the race the ability to listen to commentary through radio. Radio Le Mans is also broadcast through internet radio on their website.

Vintage racing

Since 2001, the ACO has allowed the Le Mans Legend event to participate on the full Circuit de la Sarthe during the 24 Hours week. These exhibition races involve classic cars which had previously run at Le Mans or similar to ones that had. Each year, a set era of cars is allowed to participate, with the era changing from year to year. Though mostly amateur drivers, some famous drivers have appeared to race cars they had previously run, such as Stirling Moss.

Starting in 2002, the Le Mans Classic has taken place on the full 13 km circuit in July as a biannual event. The races take places over a full 24 hour day/night cycle, with starts on set times allowing cars from the same era to compete at the same time. A team typically consists of a car in each class, and the team with the most points accumulated over five or six classes declared the overall winner. The classes are based on the era in which the cars would have competed, the exact class requirements are re-evaluated for every event since for every event the age for the youngest entries is shifted by 2 years. Although the format of the first event saw 5 classes doing more short races, later events have seen 6 classes do fewer but longer races. With the upcoming 2008 event probably allowing early Group C contenders, this format could see yet another revision with either more classes or classes spanning longer periods in time. Drivers are also required to have an FIA International Competition license to participate. This event also includes a large concours and auction.

See also

- 24 Hours of Le Mans (motorcycle race)
- 24 Hours of LeMons
- Endurance racing
- Le Mans 24 Hours video games
- List of 24 Hours of Le Mans winners
- Triple Crown of Motorsport

References

[1] "Weekly auto agenda: Le Mans" (http://www.independent.co.uk/life-style/motoring/weekly-auto-agenda-le-mans-1997970.html). The Independent. June 11, 2010. . Retrieved 22 April 2011.

[2] "Two major car families" (http://www.lemans.org/en/races/24h/discover-endurance/introduction.html). lemans.org. . Retrieved 22 April 2011.

[3] "Schedule" (http://www.lemans.org/en/races/24h/handy-information/schedule-week.html). lemans.org. . Retrieved 22 April 2011.

[4] "Le Mans 24 Hour - Michelin set new records at Le Mans" (http://uk.eurosport.yahoo.com/05082010/58/le-mans-24-hour-michelin-set-new-records-le-mans.html). Yahoo Sport/Eurosport. 23 June 2010. . Retrieved 22 April 2011.

[5] G. Harding *"A Wine Miscellany"* pg 82, Clarkson Potter Publishing, New York 2005 ISBN 0-307-34635-8

[6] "908 HY to close Le Mans Series in hybrid style". AutoBlog.

[7] "Mulsannescorner.com, "2004 Nasamax DM139"" (http://www.mulsannescorner.com/nasamaxdm139.html). Archived (http://web.archive.org/web/20080915211057/http://www.mulsannescorner.com/nasamaxdm139.html) from the original on 15 September 2008. . Retrieved 6 August 2008.

[8] "Audi R10 TDI on next generation Biofuel at Le Mans" (http://www.audiusa.com/audi/us/en2/experience/motorsport/motorsport-news/Audi_R10_TDI_uses_Biofuel_of_the_next_generation_for_the_first_time_.html). Audi Motorsport. . Retrieved 6 August 2008.

[9] "Le Mans 2009 - 2001 regulations released" (http://www.racecar-engineering.com/news/people/282760/le-mans-2009-2011-regulations-released.html). Racecar engineering. 19 November 2008. . Retrieved 11 October 2010.
[10] Dan Knutson (3 June 2003). "Points Race Stays Tight; Montoya Joins Elite Company With Victory" (http://www.usgpindy.com/news/story.php?story_id=1417). Archived (http://web.archive.org/web/20071106023236/http://www.usgpindy.com/news/story.php?story_id=1417) from the original on 6 November 2007. . Retrieved 3 December 2007.
[11] Henri Boulanger. "Monaco Grand Prix Glitz Draws Rising Stars" (http://intakeinfo.com/automotive/monaco-grand-prix-glitz-draws-rising-stars.html). IntakeInfo.com. Archived (http://web.archive.org/web/20071211040252/http://intakeinfo.com/automotive/monaco-grand-prix-glitz-draws-rising-stars.html) from the original on 11 December 2007. . Retrieved 5 December 2007.
[12] Swiss Traffic Law (German)
[13] "Eurosport" (http://tv.eurosport.co.uk/tvschedule.shtml). Archived (http://web.archive.org/web/20101013122840/http://tv.eurosport.co.uk/tvschedule.shtml) from the original on 13 October 2010. . Retrieved 11 October 2010.

- (French) Le Mans 1965 in Automobile Historique n°48 May 2005
- (French) 24 heures du Mans 1973 in Automobile Historique n°49 June/July 2005

External links

- Le Mans official website (http://www.lemans.org/en/courses/24h-du-mans.html)
- Radio Le Mans (http://www.radiolemans.com/) – Le Mans history
- Racing Sports Cars (http://www.racingsportscars.com/photo_lemans.html) – historical photos and results
- Les 24 heures (http://www.les24heures.fr/) – historical, photos and results in French

Albert_Guyot

Albert Guyot (25 December 1881 Orléans – 24 May 1947 Neuilly-sur-Seine) was a French racecar driver. He was one of four drivers who entered with Duesenberg the 1921 French Grand Prix, the first in which a US make participated. Jimmy Murphy won with his Duesenberg 183; Guyot finished 6th.

Albert Guyot

Indy 500 results

Year	Car	Start	Qual	Rank	Finish	Laps	Led	Retired
1913	9	2	80.750	19	4	200	0	Running
1914	10	11	89.150	16	3	200	9	Running
1919	32	3	98.300	9	4	200	0	Running
1921	21	15	87.780	17	8	200	0	Running
1926	39	19	88.580	27	28	8	0	Piston
Totals						808	9	

Starts	5
Poles	0
Front Row	2
Wins	0
Top 5	3
Top 10	4
Retired	1

Grand_Prix_motor_racing

Grand Prix motor racing has its roots in organised automobile racing that began in France as far back as 1894. It quickly evolved from a simple road race from one town to the next, to endurance tests for car and driver. Innovation and the drive of competition soon saw speeds exceeding 100 miles per hour (**unknown operator: u'strong'** km/h), but because the races were held on open roads there were frequent accidents with the resulting fatalities of both drivers and spectators.

Grand Prix motor racing eventually evolved into formula racing, and Formula One can be seen as its direct descendant. Each event of the Formula One World Championships is still called a Grand Prix. Formula One is still referred to as Grand Prix racing.

The origins of organised racing

Marcel Renault during the 1903 Paris Madrid trial.

Motor racing was started in France, as a direct result of the enthusiasm with which the French public embraced the motor car.[1] Manufacturers were enthusiastic due to the possibility of using motor racing as a shop window for their cars.[1] The first motor race took place on July 22, 1894 and was organised by Le Petit Journal, a Parisian newspaper. It was run over the eighty mile (128 km) distance between Paris and Rouen. The race started from Porte Maillot and went through the Bois de Boulogne. Count Jules-Albert de Dion was first into Rouen after 6 hours and 48 minutes at an average speed of 19 km/h (**unknown operator: u'strong'** mph). He finished 3'30" ahead of Georges Lemaître (Peugeot), followed by Doriot (Peugeot) at 16'30", René Panhard (Panhard) at 33'30" and Émile Levassor (Panhard) at 55'30". The official winners were Peugeot and Panhard as cars were judged on their speed, handling and safety characteristics, and De Dion's steam car needed a stoker which the judges deemed this outside of their objectives.[2] [3]

In 1900, James Gordon Bennett, Jr., the owner of the *New York Herald* newspaper and the *International Herald Tribune*, established the Gordon Bennett Cup. He hoped that the creation of an international event would drive automobile manufacturers to improve their cars.[4] Each country was allowed to enter up to three cars, which had to be fully built in the country that they represented and entered by that country's automotive governing body.[4] International racing colours were established in this event.[4] In the United States, William Kissam Vanderbilt II launched the Vanderbilt Cup at Long Island, New York in 1904.

The first Grands Prix

Georges Boillot winning the 1912 French Grand Prix in Dieppe, France

Some anglophone sources wrongly list a race called the *Pau Grand Prix* in 1901. This may stem from a mistranslation of the contemporary French sources such as the magazine *La France Auto* of March 1901.[5] The name of the 1901 event was the *Circuit du Sud-Ouest* and it was run in three classes around the streets of Pau. The *Grand Prix du Palais d'Hiver* was the name of the prizes awarded for the lesser classes ('Light cars' and 'Voiturettes'). The *Grand Prix de Pau* was the name of the prize awarded for the 'Heavy' (fastest) class. Thus Maurice Farman was awarded the 'Grand Prix de Pau' for his overall victory in the *Circuit du Sud-Ouest* driving a Panhard 24 hp. [6] [7] [8]

In *L'Histoire de l'Automobile/Paris 1907* Pierre Souvestre described the 1901 event as : "... dans le *Circuit du Sud-Ouest*, à l'occasion du meeting de Pau... " (...in the *Circuit du Sud-Ouest*, at the meeting in Pau...)[9]

The only race at the time to regularly carry the name Grand Prix was organised by the Automobile Club de France (ACF), of which the first took place in 1906. The circuit used, which was based in Le Mans, was roughly triangular in shape, each lap covering 105 kilometres (**unknown operator: u'strong'** mi). Six laps were to run each day, and each lap took approximately an hour using the relatively primitive cars of the day. The driving force behind the decision to race on a circuit - as opposed to racing on ordinary roads from town to town - was the Paris to Madrid road race of 1903. During this race a number of people, both drivers and pedestrians - including Marcel Renault - were killed and the race was stopped by the French authorities at Bordeaux. Further road based events were banned.

From the 32 entries representing 12 different automobile manufacturers, at the 1906 event, the Hungarian-born Ferenc Szisz (1873–1944) won the 1260 km (**unknown operator: u'strong'** mi) race in a Renault. This race was regarded as the first Grand Épreuve, which meant "great trial" and the term was used from then on to denote up to the eight most important events of the year.[10]

Races in this period were heavily nationalistic affairs, with a few countries setting up races of their own, but no formal championship tying them together. The rules varied from country to country and race to race, and typically centered around maximum (not minimum) weights in an effort to limit power by limiting engine size indirectly (10–15 L engines were quite common, usually with no more than four cylinders, and producing less than 50 hp). The cars all had mechanics on board as well as the driver, and no one was allowed to work on the cars during the race except for these two. A key factor to Renault winning this first Grand Prix was held to be the detachable wheel rims (developed by Michelin), which allowed tire changes to occur without having to lever the tire and tube off and back on the rim. Given the state of the roads, such repairs were frequent.

Racecourse development

For the most part, races were run over a lengthy circuit of closed public roads, not purpose-built private tracks. This was true of the Le Mans circuit of the 1906 *Grand Prix*, as well as the *Targa Florio* (run on 93 miles (**unknown operator: u'strong'** km) of Sicilian roads), the 75 miles (**unknown operator: u'strong'** km) German *Kaiserpreis* circuit in the Taunus mountains, and the French circuit at Dieppe (a mere 48 miles (**unknown operator: u'strong'** km)), used for the 1907 Grand Prix. The exceptions were the steeply banked egg-shaped near oval of Brooklands in England, completed in 1907, the Indianapolis Motor Speedway, first used in 1909 with the first Indianapolis 500-Mile Race in 1911, and the Autodromo Nazionale Monza, in Italy, opened in 1922.

In 1908, the United States of America became the first country outside France to host an automobile race using the name Grand Prix (or Grand Prize), run at Savannah. The first Grand Épreuve outside of France was the 1921 Italian

Grand Prix held at Brescia. This was quickly followed by Belgium and Spain (in 1924), and later spread to other countries including Britain (1926). Strictly speaking, this still wasn't a formal championship, but a loose collection of races run to various rules. (A "formula" of rules had appeared just before World War I, finally based on engine size as well as weight, but it was not universally adopted.)

In 1904, many national motor clubs banded together to form the Association Internationale des Automobile Clubs Reconnus (AIACR). In 1922 the Commission Sportive Internationale (CSI) was empowered on behalf of AIACR to regulate Grand Prix racing and other forms of international racing. Since the inception of Grand Prix racing, competitions had been run in accordance with a strict formula based on engine size and vehicle weight. These regulations were virtually abandoned in 1928 with an era known as *Formula Libre* when race organisers decided to run their events with almost no limitations. From 1927 to 1934, the number of races considered to have Grand Prix status exploded, jumping from five events in 1927, to nine events in 1929, to eighteen in 1934 (the peak year before World War II).

The Pre-WW II years

The 1933 Monaco Grand Prix was the first time in the history of the sport that the grid was determined by timed qualifying rather than the luck of a draw.

Grid of Coppa Fiera di Milano 1925

All the competing vehicles were painted in the international auto racing colors:

- green (British racing green) for Britain,
- blue (Bleu de France) for France,
- red (Rosso corsa) for Italian,
- yellow for Belgium, and
- white for Germany.

Beginning in 1934, the Germans stopped painting their cars, allegedly after the paint had been left off a Mercedes-Benz W25 in an effort to reduce weight. The unpainted metal soon had the German vehicles dubbed by the media as the "Silver Arrows".

1923 Karl Benz "Teardrop" racecar

French cars continued to dominate (led by Bugatti, but also including Delage and Delahaye) until the late 1920s, when the Italians (Alfa Romeo and Maserati) began to beat the French cars regularly. At the time, the Germans engineered unique race vehicles as seen in the photo here with the Benz aerodynamic "teardrop" body introduced at the 1923 European Grand Prix at Monza by Karl Benz.

In the 1930s, however, nationalism entered a new phase when the Nazis encouraged Mercedes and Auto Union to further the glory of the Reich. (The government did provide some money to the two manufacturers, but the extent of the aid into their hands was exaggerated in the media; government subsidies amounted to perhaps 10% or less of the costs of running the two racing teams.)[11] The two German marques utterly dominated the period from 1935 to 1939, winning all but three of the official Championship Grands Prix races run in those years. The cars by this time were single-seaters (the riding mechanic vanished in the early 1920s), with 8 to 16 cylinder supercharged engines producing upwards of 600 hp (**unknown operator: u'strong'** kW) on alcohol fuels.

As early as October 1923, the idea of an automobile championship was discussed at the annual autumn conference of the AIACR (Association Internationale des Automobile Clubs Reconnus) in Paris. However, discussion centered around the increased interest in racing by manufacturers and holding the first European Grand Prix at Monza in 1923. The first World Championship took place in 1925, but it was for manufacturers only, consisting of four races of at least 800 km (**unknown operator: u'strong'** mi) in length. The races that formed the first Constructors Championship were the Indianapolis 500, the European Grand Prix, and the French and Italian Grands Prix. A European Championship, consisting of the major Grand Prix in a number of countries (named *Grandes Epreuves*) was instituted for drivers in 1935, and was competed every year until the outbreak of World War II in 1939.

The post-war years and Formula One

Related topics : Formula One, History of Formula One

In 1946, following World War II, only four races of Grand Prix calibre were held. Rules for a Grand Prix World Championship had been laid out before World War II, but it took several years afterward until 1947 when the old AIACR reorganised itself as the Fédération Internationale de l'Automobile or "FIA" for short. Headquartered in Paris, at the end of the 1949 season it announced that for 1950 they would be linking several national Formula One Grands Prix to create a World Championship for drivers, although due to economic difficulties the years 1952 and 1953 were actually competed in Formula Two cars. A points system was established and a total of seven races were granted championship status including the Indianapolis 500. The first World Championship race was held on 13 May at Silverstone in the United Kingdom.

The Italians once again did well in these early World Championship races, both manufacturers and drivers. The first World Champion was Giuseppe Farina, driving an Alfa Romeo. Ferrari appeared at the second World Championship race, in Monaco, and has the distinction of being the only manufacturer to compete throughout the entire history of the World Championship, still competing in 2011.

Grandes Épreuves by season

Note : For 1950 onwards, see List of Formula One Grands Prix. Italics denote that the race was also known as the European Grand Prix.

1906–1914

Race	1906	1907	1908	1912	1913	1914
1	French	French	French	French	French	French

1921–1929

Race	1921	1922	1923	1924	1925	1926	1927	1928	1929
1	French	French	Indy 500	Indy 500	Indy 500	Indy 500	Indy 500	Indy 500	Indy 500
2	Italian	Italian	French	*French*	*Belgian*	French	French	*Italian*	French
3			*Italian*	Italian	French	*San Sebastián*	Spanish		
4					Italian	British	*Italian*		
5						Italian	British		

1930–1939

Race	1930	1931	1932	1933	1934	1935	1936	1937	1938	1939
1	Indy 500	Italian	Italian	Monaco	Monaco	Monaco	Monaco	Belgian	French	Belgian
2	*Belgian*	French	French	French	French	French	German	German	German	French
3	French	Belgian	German	Belgian	German	Belgian	Swiss	Monaco	Swiss	German
4		German		Italian	Belgian	German	Italian	Swiss	Italian	Swiss
5				Spanish	Italian	Swiss		Italian		
6					Spanish	Italian				
7						Spanish				

For wartime events, see 1940-1945 Grand Prix season.

1946–1949

Race	1946	1947	1948	1949
1	Saint-Cloud	Swiss	Monaco	British
2	Nations	*Belgian*	*Swiss*	Belgian
3	Turin	Italian	French	Swiss
4		French	Italian	French
5			British	*Italian*

Other events included

- Argentine Grand Prix
- Avus Grand Prix
- Bari Grand Prix
- Belgrade Grand Prix
- Buenos Aires Grand Prix
- Coppa Acerbo
- Coppa Ciano
- Czechoslovakian Grand Prix
- Donington Grand Prix
- Dutch Grand Prix
- Hungarian Grand Prix
- Lviv Grand Prix
- Milan Grand Prix
- Mille Miglia
- Moroccan Grand Prix
- New Zealand Grand Prix
- Penya Rhin Grand Prix
- Russian Grand Prix
- San Sebastián Grand Prix

- Stockholm Grand Prix (Formula Two)
- Swedish Summer Grand Prix
- Targa Florio
- Tripoli Grand Prix
- Tunis Grand Prix
- United States Grand Prix
- Vanderbilt Cup
- Zandvoort Grand Prix

See also:

- List of major automobile races in France
- List of major automobile races in Germany
- List of major automobile races in Italy
- List of major automobile races in Spain

Grand Prix drivers

Notable drivers of the Grand Prix motor racing era included a few women who competed equally with the men:

- Antonio Ascari - Italy
- Robert Benoist - France
- Clemente Biondetti - Italy
- Georges Boillot - France
- Manfred von Brauchitsch - Germany
- Malcolm Campbell - Great Britain
- Rudolf Caracciola - Germany
- / Luigi Chinetti - Italy; United States after the war
- Louis Chiron - Monaco
- Albert Divo - France
- René Dreyfus - France
- Philippe Étancelin - France
- Luigi Fagioli - Italy
- Giuseppe Farina - Italy; he became the first Formula One champion
- Enzo Ferrari - Italy
- Jules Goux - France
- Elizabeth Junek - Czechoslovakia
- Hermann Lang - Germany
- Christian Lautenschlager - Germany
- Emilio Materassi - Italy
- Felice Nazzaro - Italy
- Guy Moll - Algeria
- Hellé Nice - France
- Tazio Nuvolari - Italy
- Kay Petre - Great Britain
- Charles Pozzi - France
- Georges Philippe (Baron Philippe de Rothschild) - France
- Bernd Rosemeyer - Germany
- Richard Seaman - Great Britain
- Henry Segrave - Great Britain

- Raymond Sommer - France
- Whitney Willard Straight - Great Britain
- Hans Stuck - Germany
- Ferenc Szisz - Hungary
- Achille Varzi - Italy
- Emilio Villoresi - Italy
- Luigi Villoresi - Italy
- William Grover-Williams - Great Britain
- Jean-Pierre Wimille - France
- Juan Zanelli - Chile

Championships

From 1925 onwards, the AIACR and later the FIA organised World and European Championships for Grand Prix manufacturers, drivers and constructors:

- World Manufacturers' Championship (1925-1927)
- European Drivers' Championship (1931-1932, 1935-1939)
- World Drivers' Championship (1950-1980)
- International Cup for Constructors (1958-1980)
- Formula One World Drivers' Championship (1981-present)
- Formula One World Constructors' Championship (1981-present)

Notes

[1] Rendall, Ivan (1995). *The Chequered Flag*. Weidenfeld and Nicolson. p. 10. ISBN 0-297-83550-5.

[2] "Forix, Autosport, 8W - Welcome to Who? What? Where? When? Why? on the World Wide Web. "The cradle of motorsport" by Rémi Paolozzi, May 28, 2003" (http://forix.autosport.com/8w/bdb.html). Forix.autosport.com. . Retrieved 2011-09-24.

[3] Rendall, Ivan (1995). *The Chequered Flag*. Weidenfeld and Nicolson. p. 12. ISBN 0-297-83550-5.

[4] Rendall, Ivan (1995). *The Chequered Flag*. Weidenfeld and Nicolson. p. 26. ISBN 0-297-83550-5.

[5] Autosport, The Nostalgia Forum, Robert Dick quoting (translating) from *La France Automobile*, March 1901 (http://forums.autosport.com/lofiversion/index.php/t57839.html)

[6] Racing within the city of Pau dates from 1900 when the first edition of the *Circuit du Sud-Ouest* was run in the city.

- The first *Circuit du Sud-Ouest* on 25 February 1900 was won by René de Knyff in a Panhard 16 hp. (*Grand Prix Winners 1895–1949* by Hans Etzrodt Part 1 1895–1916 (http://www.kolumbus.fi/leif.snellman/gpw1.htm))

[7] *La France Automobile*, March 1901 reports the results for the 'Semaine de Pau'. There were two discrete events, the 140 km *Course des touristes* from Pau-Peyrehorade-Pau and the second edition of the *Circuit du Sud-Ouest* on 17 February 1901 around a course on the city outskirts.

The *Course des touristes* comprised six prizes for the different classes of entrants. The *Prix de la Presse* was won by Barbereau (De Dietrich); the *Prix du Commerce Palois* was awarded to Henri Farman (Darracq); the *Prix des Cercles* was awarded to Rudeaux (Darracq); the *Prix de l'Automobile Club Béarnais* was won by 'Bergeon' (De Dietrich); the *Prix de l'Automobile Club de France* was awarded to Demeester (Gladiator) and Edmond (Darracq); the *Prix de Palmarium* was awarded to Cormier (De Dion).(Autosport, The Nostalgia Forum, Robert Dick quoting (translating) from *La France Automobile*, March 1901 (http:/ / forums. autosport. com/ lofiversion/ index. php/ t57839.html)

The *Circuit du Sud-Ouest* comprised four prizes for the different classes of entrants.

- The *Grand Prix de Pau* was awarded to Maurice Farman (Panhard 24 hp).
- The *Grand Prix du Palais d'Hiver* (400 à 650 kg 'Light car' class) was awarded to Henri Farman (Darracq).
- The second *Grand Prix du Palais d'Hiver* for the under 400 kg *Voiturettes*class was awarded to Louis Renault (Renault).
- The *Prix du Béarn* was awarded to Osmont in a 'De Dion' tricycle.

((*Grand Prix Winners 1895–1949* by Hans Etzrodt Part 1 1895–1916 (http://www.kolumbus.fi/leif.snellman/gpw1.htm))

[8] By the turn of the century the term *Grand Prix* had become common parlance in France, having been used since the Grand Prix de Paris horse race in 1886 (e.g. the Grand Prix de Paris for Cyclists in 1895) (New York Times, July 18, 1895, Grand Prix de Paris for Cyclists (http://query.nytimes.com/mem/archive-free/pdf?res=FA0A11FB3B5911738DDDA10994DF405B8585F0D3)). Thus in the anglophone world the main winner's prize (*Grand Prix de Pau*) subsequently became synonymous with the event.(Leif Snellman (2002-05-27). "The first Grand Prix" (http://forix.autosport.com/8w/f06.html). *8W*. FORIX. . Retrieved 2011-01-28.)

[9] Autosport, The Nostalgia Forum, Robert Dick quoting (translating) from *'L'Histoire de l'Automobile* by Pierre Souvestre, published in Paris 1907 (http://forums.autosport.com/lofiversion/index.php/t57839.html)

[10] Etzrodt, Hans. "Grand Prix Winners 1895-1949" (http://www.kolumbus.fi/leif.snellman/gpw4.htm#GRA). The Golden Age. . Retrieved 2007-04-03.

[11] Nixon, Chris (1986). *Racing the Silver Arrows*. Osprey Publishing Limited. p. 155. ISBN 0-85184-055-8.

See also

- Formula Grand Prix

References

External links

- Grandprix-Live.com (http://www.grandprix-live.com/) - Covers all the major Grand Prix motorsport championships, including Formula One, MotoGP, NASCAR, IRL, WSBK, WRC Rally and many more.
- Grand Prix History (http://www.ddavid.com/formula1/story.htm) - The Story of the Grand Prix

World_War_I

World War I (**WWI**), which was predominantly called the **World War** or the **Great War** from its occurrence until 1939, and the **First World War** or World War I thereafter, was a major war centred in Europe that began on 28 July 1914 and lasted until 11 November 1918. It involved all the world's great powers,[1] which were assembled in two opposing alliances: the Allies (based on the Triple Entente of the United Kingdom, France and Russia) and the Central Powers (originally centred around the Triple Alliance of Germany, Austria-Hungary and Italy; but, as Austria–Hungary had taken the offensive against the agreement, Italy did not enter into the war).[2] These alliances both reorganised (Italy fought for the Allies), and expanded as more nations entered the war. Ultimately more than 70 million military personnel, including 60 million Europeans, were mobilised in one of the largest wars in history.[3] [4] More than 9 million combatants were killed, largely because of great technological advances in firepower without corresponding advances in mobility. It was the sixth-deadliest conflict in world history, subsequently paving the way for various political changes such as revolutions in the nations involved.[5]

Long-term causes of the war included the imperialistic foreign policies of the great powers of Europe, including the German Empire, the Austro-Hungarian Empire, the Ottoman Empire, the Russian Empire, the British Empire, the French Republic, and Italy. The assassination on 28 June 1914 of Archduke Franz Ferdinand of Austria, the heir to the throne of Austria-Hungary, by a Yugoslav nationalist was the proximate trigger of the war. It resulted in a Habsburg ultimatum against the Kingdom of Serbia.[6] [7] Several alliances formed over the previous decades were invoked, so within weeks the major powers were at war; via their colonies, the conflict soon spread around the world.

On 28 July, the conflict opened with the Austro-Hungarian invasion of Serbia,[8] [9] followed by the German invasion of Belgium, Luxembourg and France; and a Russian attack against Germany. After the German march on Paris was brought to a halt, the Western Front settled into a static battle of attrition with a trench line that changed little until 1917. In the East, the Russian army successfully fought against the Austro-Hungarian forces but was forced back by the German army. Additional fronts opened after the Ottoman Empire joined the war in 1914, Italy and Bulgaria in 1915 and Romania in 1916. The Russian Empire collapsed in March 1917, and Russia left the war after the October Revolution later that year. After a 1918 German offensive along the western front, United States forces entered the trenches and the Allies drove back the German armies in a series of successful offensives. Germany, which had its own trouble with revolutionaries at this point, agreed to a cease-fire on 11 November 1918, later known as Armistice Day. The war had ended in victory for the Allies.

Events on the home fronts were as tumultuous as on the battle fronts, as the participants tried to mobilize their manpower and economic resources to fight a total war. By the end of the war, four major imperial powers —the German, Russian, Austro-Hungarian and Ottoman empires— ceased to exist. The successor states of the former two lost a great amount of territory, while the latter two were dismantled entirely. The map of central Europe was redrawn into several smaller states.[10] The League of Nations was formed in the hope of preventing another such conflict. The European nationalism spawned by the war and the breakup of empires, the repercussions of Germany's defeat and problems with the Treaty of Versailles are generally agreed to be factors contributing to World War II.[11]

Names

In Canada, *Maclean's Magazine* in October 1914 said, "Some wars name themselves. This is the Great War."[12] During the Interwar period, the war was most often called the *World War* and the *Great War*, in English-speaking countries.

After the onset of the Second World War in 1939, the terms *World War I* or *the First World War* became standard, with British and Canadian historians favouring the *First World War*, and Americans *World War I*. The notion that the "World War" was merely the first in a series was not a new idea at the time, however; it was first introduced in September 1914 by German philosopher Ernst Haeckel, who claimed that "there is no doubt that the course and character of the feared 'European War' ... will become the first world war in the full sense of the word."[13] *The First World War* was the title of a 1920 history by the officer and journalist Charles à Court Repington.

Background

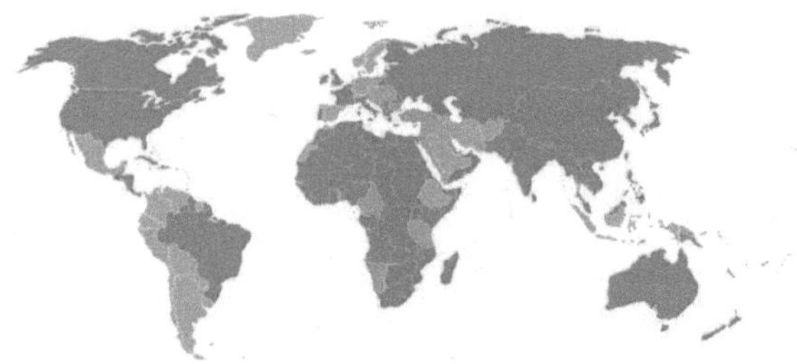

Map of the participants in World War I: Allied Powers in green, Central Powers in orange, and neutral countries in grey

In the 19th century, the major European powers had gone to great lengths to maintain a balance of power throughout Europe, resulting by 1900 in a complex network of political and military alliances throughout the continent.[2] These had started in 1815, with the Holy Alliance between Prussia, Russia, and Austria. Then, in October 1873, German Chancellor Bismarck negotiated the League of the Three Emperors (German: *Dreikaiserbund*) between the monarchs of Austria–Hungary, Russia and Germany. This agreement failed because Austria–Hungary and Russia could not agree over Balkan policy, leaving Germany and Austria–Hungary in an alliance formed in 1879, called the Dual Alliance. This was seen as a method of countering Russian influence in the Balkans as the Ottoman Empire continued to weaken.[2] In 1882, this alliance was expanded to include Italy in what became the Triple Alliance.[14]

After 1870, European conflict was averted largely through a carefully planned network of treaties between the German Empire and the remainder of Europe orchestrated by Bismarck. He especially worked to hold Russia at

Germany's side to avoid a two-front war with France and Russia. When Wilhelm II ascended to the throne as German Emperor (*Kaiser*), Bismarck's alliances were gradually de-emphasised. For example, the Kaiser refused to renew the Reinsurance Treaty with Russia in 1890. Two years later, the Franco-Russian Alliance was signed to counteract the force of the Triple Alliance. In 1904, the United Kingdom sealed an alliance with France, the Entente Cordiale, and in 1907, the United Kingdom and Russia signed the Anglo-Russian Convention. This system of interlocking bilateral agreements formed the Triple Entente.[2]

HMS *Dreadnought*. A naval arms race existed between the United Kingdom and Germany.

German industrial and economic power had grown greatly after unification and the foundation of the Empire in 1870. From the mid-1890s on, the government of Wilhelm II used this base to devote significant economic resources to building up the *Kaiserliche Marine* (Imperial German Navy), established by Admiral Alfred von Tirpitz, in rivalry with the British Royal Navy for world naval supremacy.[15] As a result, each nation strove to out-build the other in terms of capital ships. With the launch of HMS *Dreadnought* in 1906, the British Empire expanded on its significant advantage over its German rival.[15] The arms race between Britain and Germany eventually extended to the rest of Europe, with all the major powers devoting their industrial base to producing the equipment and weapons necessary for a pan-European conflict.[16] Between 1908 and 1913, the military spending of the European powers increased by 50 percent.[17]

Gavrilo Princip, a Bosnian-Serb student, was arrested immediately after he assassinated Archduke Franz Ferdinand of Austria

Austria-Hungary precipitated the Bosnian crisis of 1908–1909 by officially annexing the former Ottoman territory of Bosnia and Herzegovina, which it had occupied since 1878. This angered the Kingdom of Serbia and its patron, the Pan-Slavic and Orthodox Russian Empire.[18] Russian political manoeuvring in the region destabilised peace accords that were already fracturing in what was known as "the powder keg of Europe".[18]

In 1912 and 1913 the First Balkan War was fought between the Balkan League and the fracturing Ottoman Empire. The resulting Treaty of London further shrank the Ottoman Empire, creating an independent Albanian State while enlarging the territorial holdings of Bulgaria, Serbia, Montenegro, and Greece. When Bulgaria attacked both Serbia and Greece on 16 June 1913, it lost most of Macedonia to Serbia and Greece and Southern Dobruja to Romania in the 33-day Second Balkan War, further destabilising the region.[19]

Ethno-linguistic map of Austria–Hungary, 1910

On 28 June 1914, Gavrilo Princip, a Bosnian-Serb student and member of Young Bosnia, assassinated the heir to the Austro-Hungarian throne, Archduke Franz Ferdinand of Austria in Sarajevo, Bosnia.[20] This began a period of diplomatic manoeuvring among Austria-Hungary, Germany, Russia, France, and Britain called the July Crisis. Wanting to finally end Serbian interference in Bosnia, Austria-Hungary delivered the July Ultimatum to Serbia, a series of ten demands intentionally made unacceptable, intending to provoke a war with Serbia.[21] When Serbia agreed to only eight of the ten demands, Austria-Hungary declared war on 28 July 1914. Strachan argues, "Whether an equivocal and early response by Serbia would have made any difference to Austria-Hungary's behaviour must be doubtful. Franz Ferdinand was not the sort of personality who commanded popularity, and his demise did not cast the empire into deepest mourning".[22]

The Russian Empire, unwilling to allow Austria–Hungary to eliminate its influence in the Balkans, and in support of its longtime Serb protégés, ordered a partial mobilisation one day later.[14] When the German Empire began to mobilise on 30 July 1914, France, resentful of the German conquest of Alsace-Lorraine during the Franco-Prussian War, ordered French mobilisation on 1 August. Germany declared war on Russia on the same day.[23] The United Kingdom declared war on Germany on 4 August 1914, following an "unsatisfactory reply" to the British ultimatum that Belgium must be kept neutral.[24]

Theatres of Conflict

Opening hostilities

Confusion among the Central Powers

The strategy of the Central Powers suffered from miscommunication. Germany had promised to support Austria-Hungary's invasion of Serbia, but interpretations of what this meant differed. Previously-tested deployment plans had been replaced early in 1914, but the replacements had never been tested in exercises. Austro-Hungarian leaders believed Germany would cover its northern flank against Russia.[25] Germany, however, envisioned Austria-Hungary directing most of its troops against Russia, while Germany dealt with France. This confusion forced the Austro-Hungarian Army to divide its forces between the Russian and Serbian fronts.

On 9 September 1914, the *Septemberprogramm*, a possible plan which detailed Germany's specific war aims and the conditions that Germany sought to force on the Allied Powers, was outlined by German Chancellor Theobald von Bethmann-Hollweg. It was never officially adopted.

African campaigns

Lettow surrendering his forces to the British at Abercorn

Some of the first clashes of the war involved British, French, and German colonial forces in Africa. On 7 August, French and British troops invaded the German protectorate of Togoland. On 10 August, German forces in South-West Africa attacked South Africa; sporadic and fierce fighting continued for the rest of the war. The German colonial forces in German East Africa, led by Colonel Paul Emil von Lettow-Vorbeck, fought a guerrilla warfare campaign during World War I and only surrendered two weeks after the armistice took effect in Europe.[26]

Serbian campaign

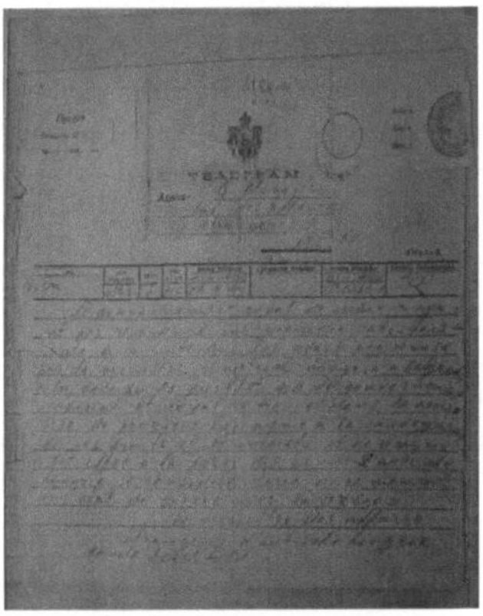

Declaration of war. Austro-Hungarian government's telegram to the government of Serbia on 28 July 1914.

Austria invaded and fought the Serbian army at the Battle of Cer and Battle of Kolubara beginning on 12 August. Over the next two weeks Austrian attacks were thrown back with heavy losses, which marked the first major Allied victories of the war and dashed Austro-Hungarian hopes of a swift victory. As a result, Austria had to keep sizable forces on the Serbian front, weakening its efforts against Russia.[27] Serbia's defeat of the Austro-Hungarian invasion of 1914 counts among the major upset victories of the last century.[28]

Serbian troops artillery positions in the Battle of Kolubara.

German forces in Belgium and France

German soldiers in a railway goods van on the way to the front in 1914. A message on the car spells out "Trip to Paris"; early in the war all sides expected the conflict to be a short one.

At the outbreak of the First World War, the German army (consisting in the West of seven field armies) carried out a modified version of the Schlieffen Plan, designed to quickly attack France through neutral Belgium before turning southwards to encircle the French army on the German border.[6] The plan called for the right flank of the German advance to converge on Paris, and initially the Germans were successful, particularly in the Battle of the Frontiers (14–24 August). By 12 September, the French, with assistance from the British forces, halted the German advance east of Paris at the First Battle of the Marne (5–12 September). The last days of this battle signified the end of mobile warfare in the west.[6] The French offensive into Germany, launched on 7 August with the Battle of Mulhouse, had limited success.

In the east, only one field army defended East Prussia, and when Russia attacked in this region it diverted German forces intended for the Western Front. Germany defeated Russia in a series of battles collectively known as the First Battle of Tannenberg (17 August – 2 September), but this diversion aggravated problems of insufficient speed of advance from rail-heads not foreseen by the German General Staff. The Central Powers were denied a quick victory and forced to fight a war on two fronts. The German army had fought its way into a good defensive position inside France and had permanently incapacitated 230,000 more French and British troops than it had lost itself. Despite this, communications problems and questionable command decisions cost Germany the chance of early victory.[29]

Asia and the Pacific

New Zealand occupied German Samoa (later Western Samoa) on 30 August. On 11 September, the Australian Naval and Military Expeditionary Force landed on the island of Neu Pommern (later New Britain), which formed part of German New Guinea. Japan seized Germany's Micronesian colonies and, after the Siege of Tsingtao, the German coaling port of Qingdao in the Chinese Shandong peninsula. Within a few months, the Allied forces had seized all the German territories in the Pacific; only isolated commerce raiders and a few holdouts in New Guinea remained.[30] [31]

Early stages

Trench warfare begins

Military tactics before World War I had failed to keep pace with advances in technology. These advances allowed for impressive defence systems, which out-of-date military tactics could not break through for most of the war. Barbed wire was a significant hindrance to massed infantry advances. Artillery, vastly more lethal than in the 1870s, coupled with machine guns, made crossing open ground extremely difficult.[32] The Germans introduced poison gas; it soon became used by both sides, though it never proved decisive in winning a battle. Its effects were brutal, causing slow and painful death, and poison gas became one of the most-feared and best-remembered horrors of the war. Commanders on both sides failed to develop tactics for breaching entrenched positions without heavy casualties.

Sir Winston Churchill with the Royal Scots Fusiliers, 1916

In time, however, technology began to produce new offensive weapons, such as the tank.[33] Britain and France were its primary users; the Germans employed captured Allied tanks and small numbers of their own design. After the First Battle of the Marne, both Entente and German forces began a series of outflanking manoeuvres, in the so-called "Race to the Sea". Britain and France soon found themselves facing entrenched German forces from Lorraine to Belgium's coast.[6] Britain and France sought to take the offensive, while Germany defended the occupied territories; consequently, German trenches were much better constructed than those of their enemy. Anglo-French trenches were only intended to be "temporary" before their forces broke through German defences.[34] Both sides tried to break the stalemate using scientific and technological advances. On 22 April 1915 at the Second Battle of Ypres, the Germans (violating the Hague Convention) used chlorine gas for the first time on the Western Front. Algerian troops retreated when gassed and a six-kilometre (four-mile) hole opened in the Allied lines that the Germans quickly exploited, taking Kitcheners' Wood. Canadian soldiers closed the breach at the Second Battle of Ypres.[35] At the Third Battle of Ypres, Canadian and ANZAC troops took the village of Passchendaele.

Sunlight Soap ad, placed in a trench (1915)

On 1 July 1916, the British Army endured the bloodiest day in its history, suffering 57,470 casualties, including 19,240 dead, on the first day of the Battle of the Somme. Most of the casualties occurred in the first hour of the attack. The entire Somme offensive cost the British Army almost half a million men.[36]

Men in Melbourne collecting recruitment papers

Neither side proved able to deliver a decisive blow for the next two years, though protracted German action at Verdun throughout 1916,[37] combined with the bloodletting at the Somme, brought the exhausted French army to the brink of collapse. Futile attempts at frontal assault came at a high price for both the British and the French *poilu* (infantry) and led to widespread mutinies, especially during the Nivelle Offensive.[38]

In the trenches: Royal Irish Rifles in a communications trench on the first day on the Somme, 1 July 1916.

Canadian troops advancing behind a British Mark II tank at the Battle of Vimy Ridge.

Throughout 1915–17, the British Empire and France suffered more casualties than Germany, because of both the strategic and tactical stances chosen by the sides. Strategically, while the Germans only mounted a single main offensive at Verdun, the Allies made several attempts to break through German lines. Tactically, German commander Erich Ludendorff's doctrine of "elastic defence" was well suited for trench warfare. This defence had a lightly defended forward position and a more powerful main position farther back beyond artillery range, from which an immediate and powerful counter-offensive could be launched.[39] [40]

Ludendorff wrote on the fighting in 1917,

> The 25th of August concluded the second phase of the Flanders battle. It had cost us heavily ... The costly August battles in Flanders and at Verdun imposed a heavy strain on the Western troops. In spite of all the concrete protection they seemed more or less powerless under the enormous weight of the enemy's artillery. At some points they no longer displayed the firmness which I, in common with the local commanders, had hoped for. The enemy managed to adapt himself to our method of employing counter attacks ... I myself was being put to a terrible strain. The state of affairs in the West appeared to prevent the execution of our plans elsewhere. Our wastage had been so high as to cause grave misgivings, and had exceeded all expectation.[41]

A French assault on German positions. Champagne, France, 1917.

On the battle of the Menin Road Ridge, Ludendorff wrote,

> Another terrific assault was made on our lines on the 20 September ... The enemy's onslaught on the 20th was successful, which proved the superiority of the attack over the defence. Its strength did not consist in the tanks; we found them inconvenient, but put them out of action all the same. The power of the attack lay in the artillery, and in the fact that ours did not do enough damage to the hostile infantry as they were assembling, and above all, at the actual time of the assault.[42]

Officers and senior enlisted men of the Bermuda Militia Artillery's Bermuda Contingent, Royal Garrison Artillery, in Europe.

Around 1.1 to 1.2 million soldiers from the British and Dominion armies were on the Western Front at any one time.[43] A thousand battalions, occupying sectors of the line from the North Sea to the Orne River, operated on a month-long four-stage rotation system, unless an offensive was underway. The front contained over 9600 kilometres (**unknown operator: u'strong'** mi) of trenches. Each battalion held its sector for about a week before moving back to support lines and then further back to the reserve lines before a week out-of-line, often in the Poperinge or Amiens areas.

The British Grand Fleet making steam for Scapa Flow, 1914

In the 1917 Battle of Arras, the only significant British military success was the capture of Vimy Ridge by the Canadian Corps under Sir Arthur Currie and Julian Byng. The assaulting troops could, for the first time, overrun, rapidly reinforce and hold the ridge defending the coal-rich Douai plain.[44] [45]

A battleship squadron of the Hochseeflotte at sea

Naval war

At the start of the war, the German Empire had cruisers scattered across the globe, some of which were subsequently used to attack Allied merchant shipping. The British Royal Navy systematically hunted them down, though not without some embarrassment from its inability to protect Allied shipping. For example, the German detached light cruiser SMS *Emden*, part of the East-Asia squadron stationed at Tsingtao, seized or destroyed 15 merchantmen, as well as sinking a Russian cruiser and a French destroyer. However, most of the German East-Asia squadron—consisting of the armoured cruisers *Scharnhorst* and *Gneisenau*, light cruisers *Nürnberg* and *Leipzig* and two transport ships—did not have orders to raid shipping and was instead underway to Germany when it met British warships. The German flotilla and *Dresden* sank two armoured cruisers at the Battle of Coronel, but was almost destroyed at the Battle of the Falkland Islands in December 1914, with only *Dresden* and a few auxiliaries escaping, but at the Battle of Más a Tierra these too were destroyed or interned.[46]

Soon after the outbreak of hostilities, Britain began a naval blockade of Germany. The strategy proved effective, cutting off vital military and civilian supplies, although this blockade violated accepted international law codified by several international agreements of the past two centuries.[47] Britain mined international waters to prevent any ships from entering entire sections of ocean, causing danger to even neutral ships.[48] Since there was limited response to this tactic, Germany expected a similar response to its unrestricted submarine warfare.[49]

The 1916 Battle of Jutland (German: *Skagerrakschlacht*, or "Battle of the Skagerrak") developed into the largest naval battle of the war, the only full-scale clash of battleships during the war, and one of the largest in history. It took place on 31 May – 1 June 1916, in the North Sea off Jutland. The Kaiserliche Marine's High Seas Fleet, commanded by Vice Admiral Reinhard Scheer, squared off against the Royal Navy's Grand Fleet, led by Admiral Sir John Jellicoe. The engagement was a stand off, as the Germans, outmanoeuvred by the larger British fleet, managed to escape and inflicted more damage to the British fleet than they received. Strategically, however, the British asserted their control of the sea, and the bulk of the German surface fleet remained confined to port for the duration of the war.[50]

German U-boats attempted to cut the supply lines between North America and Britain.[51] The nature of submarine warfare meant that attacks often came without warning, giving the crews of the merchant ships little hope of survival.[51] [52] The United States launched a protest, and Germany changed its rules of engagement. After the notorious sinking of the passenger ship RMS *Lusitania* in 1915, Germany promised not to target passenger liners, while Britain armed its merchant ships, placing them beyond the protection of the "cruiser rules" which demanded warning and placing crews in "a place of safety" (a standard which lifeboats did not meet).[53] Finally, in early 1917

Germany adopted a policy of unrestricted submarine warfare, realising the Americans would eventually enter the war.[51] [54] Germany sought to strangle Allied sea lanes before the U.S. could transport a large army overseas, but could maintain only five long-range U-boats on station, to limited effect.[51]

The U-boat threat lessened in 1917, when merchant ships began travelling in convoys, escorted by destroyers. This tactic made it difficult for U-boats to find targets, which significantly lessened losses; after the hydrophone and depth charges were introduced, accompanying destroyers might attack a submerged submarine with some hope of success. Convoys slowed the flow of supplies, since ships had to wait as convoys were assembled. The solution to the delays was an extensive program to build new freighters. Troopships were too fast for the submarines and did not travel the North Atlantic in convoys.[55] The U-boats had sunk more than 5,000 Allied ships, at a cost of 199 submarines.[56]

U-155 exhibited near Tower Bridge in London after the First World War.

World War I also saw the first use of aircraft carriers in combat, with HMS *Furious* launching Sopwith Camels in a successful raid against the Zeppelin hangars at Tondern in July 1918, as well as blimps for antisubmarine patrol.[57]

Southern theatres

War in the Balkans

Faced with Russia, Austria-Hungary could spare only one-third of its army to attack Serbia. After suffering heavy losses, the Austrians briefly occupied the Serbian capital, Belgrade. A Serbian counter attack in the battle of Kolubara, however, succeeded in driving them from the country by the end of 1914. For the first ten months of 1915, Austria-Hungary used most of its military reserves to fight Italy. German and Austro-Hungarian diplomats, however, scored a coup by persuading Bulgaria to join in attacking Serbia. The Austro-Hungarian provinces of Slovenia, Croatia and Bosnia provided troops for Austria-Hungary, invading Serbia as well as fighting Russia and Italy. Montenegro allied itself with Serbia.[58]

Serbia was conquered in a little more than a month, as the Central Powers, now including Bulgaria, sent in 600,000 troops. The Serbian army, fighting on two fronts and facing certain defeat, retreated into Albania, The Serbs suffered defeat in the Battle of Kosovo. Montenegro covered the Serbian retreat towards the Adriatic coast in the Battle of Mojkovac in 6–7 January 1916, but ultimately the Austrians conquered Montenegro, too. The surviving 70,000 Serbian soldiers were evacuated by ship to Greece.[59]

In late 1915, a Franco-British force landed at Salonica in Greece, to offer assistance and to pressure the government to declare war against the Central Powers. Unfortunately for the Allies, the pro-German King Constantine I dismissed the pro-Allied government of Eleftherios Venizelos, before the Allied expeditionary force could arrive.[60] The friction between the king of Greece and the Allies continued to accumulate with the National Schism, which effectively divided Greece between regions still loyal to the king and the new provisional government of Venizelos in Salonica. After intensive diplomatic negotiations and an armed confrontation in Athens between Allied and royalist forces (an incident known as Noemvriana) the king of Greece resigned, and his second son Alexander took his place. Venizelos returned to Athens on 29 May 1917 and Greece, now unified, officially joined the war on the side of the Allies. The entire Greek army was mobilized and began to participate in military operations against the Central Powers on the Macedonian front.

After conquest, Serbia was divided between Austro-Hungary and Bulgaria. In 1917 the Serbs launched the Toplica Uprising and liberated for a short time the area between the Kopaonik mountains and the South Morava river. The uprising was crushed by joint efforts of Bulgarian and Austrian forces at the end of March 1917.

Bulgarian soldiers in a trench, preparing to fire against an incoming airplane

The Macedonian Front in the beginning was mostly static. French and Serbian forces retook limited areas of Macedonia by recapturing Bitola on 19 November 1916 as a result of the costly Monastir Offensive which brought stabilization of the front.

Serbian and French troops finally made a breakthrough, after most of the German and Austro-Hungarian troops had withdrawn. This breakthrough was significant in defeating Bulgaria and Austro-Hungary, which led to the final victory of WWI. The Bulgarians suffered their only defeat of the war at the Battle of Dobro Pole but days later, they decisively defeated British and Greek forces at the Battle of Doiran, avoiding occupation. After Serbian breakthrough of Bulgarian lines, Bulgaria capitulated on 29 September 1918.[61] Hindenburg and Ludendorff concluded that the strategic and operational balance had now shifted decidedly against the Central Powers and a day after the Bulgarian collapse, during a meeting with government officials, insisted on an immediate peace settlement.[62]

The disappearance of the Macedonian front meant that the road to Budapest and Vienna was now opened for the 670,000-strong army of general Franchet d'Esperey as the Bulgarian surrender deprived the Central Powers of the 278 infantry battalions and 1,500 guns (the equivalent of some 25 to 30 German divisions) that were previously holding the line.[63] The German high command responded by sending only seven infantry and one cavalry division but these forces were far from enough for a front to be reestablished.[63]

Ottoman Empire

The Ottoman Empire joined the Central Powers in the war, the secret Ottoman-German Alliance having been signed in August 1914.[64] It threatened Russia's Caucasian territories and Britain's communications with India via the Suez Canal. The British and French opened overseas fronts with the Gallipoli (1915) and Mesopotamian campaigns. In Gallipoli, the Ottoman Empire successfully repelled the British, French, and Australian and New Zealand Army Corps (ANZACs). In Mesopotamia, by contrast, after the disastrous Siege of Kut (1915–16), British Imperial forces reorganised and captured Baghdad in March 1917.

A British artillery battery emplaced on Mount Scopus in the Battle of Jerusalem.

Further to the west, the Suez Canal was successfully defended from Ottoman attacks in 1915 and 1916; in August a joint German and Ottoman force was defeated at the Battle of Romani by the Anzac Mounted and the 52nd (Lowland) Infantry Divisions. Following this victory, a British Empire Egyptian Expeditionary Force advanced across the Sinai Peninsula, pushing Ottoman forces back in the Battle of Magdhaba in December and the Battle of Rafa on the border between the Egyptian Sinai and Ottoman Palestine in January 1917. In March and April at the First and Second Battles of Gaza, German and Ottoman forces stopped the advance, but at the end of October the Sinai and Palestine Campaign resumed, when Allenby's XXth Corps, XX1st Corps and Desert Mounted Corps won the Battle of Beersheba. Two Ottoman armies were defeated a few weeks later at the Battle of Mughar Ridge, and early in December Jerusalem was captured following another Ottoman defeat at the Battle of Jerusalem (1917). About this time Friedrich Freiherr Kress von Kressenstein was relieved of his duties as the Eighth Army's commander, replaced by Djevad Pasha, and a few months later the commander of the Ottoman Army in Palestine, Erich von Falkenhayn, was replaced by Otto Liman von Sanders.

A reorganised Egyptian Expeditionary Force, with an additional mounted division under Field Marshal Edmund Allenby, broke Ottoman forces at the Battle of Megiddo in September 1918. In six weeks, during virtually continuous operations, battles were successfully fought by British infantry and Australian, British, Indian and New Zealand, Light Horse, mounted Yeomanry, Lancers and Mounted Rifle brigades. They campaigned across the Jordan River to Amman in the east and northwards to capture Nablus and Tulkarm in the Judean Hills, and followed the Mediterranean coast into the Jezreel Valley (Esdraelon Plain), where Afula, Jenin and Nazareth were captured, along with Daraa east of the Jordan River on the Hejaz railway. Semakh and Tiberias on the Sea of Gallilee, were captured on the way northwards to Damascus and Aleppo. The Armistice of Mudros was signed on 30 October, when two and a half Ottoman armies had been defeated and captured.

Russian forest trench at the Battle of Sarikamish

Russian armies generally had the best of it in the Caucasus. Enver Pasha, supreme commander of the Ottoman armed forces, was ambitious and dreamed of re-conquering central Asia and areas that had been lost to Russia previously. He was, however, a poor commander.[65] He launched an offensive against the Russians in the Caucasus in December 1914 with 100,000 troops; insisting on a frontal attack against mountainous Russian positions in winter, he lost 86% of his force at the Battle of Sarikamish.[66]

General Yudenich, the Russian commander from 1915 to 1916, drove the Turks out of most of the southern Caucasus with a string of victories.[66] In 1917, Russian Grand Duke Nicholas assumed command of the Caucasus front. Nicholas planned a railway from Russian Georgia to the conquered territories, so that fresh supplies could be brought up for a new offensive in 1917. However, in March 1917 (February in the pre-revolutionary Russian calendar), the Czar was overthrown in the February Revolution and the Russian Caucasus Army began to fall apart.

German soldiers in Jerusalem

Instigated by the Arab bureau of the British Foreign Office, the Arab Revolt started with the help of Britain in June 1916 at the Battle of Mecca, led by Sherif Hussein of Mecca, and ended with the Ottoman surrender of Damascus. Fakhri Pasha, the Ottoman commander of Medina, resisted for more than two and half years during the Siege of Medina.[67]

Along the border of Italian Libya and British Egypt, the Senussi tribe, incited and armed by the Turks, waged a small-scale guerrilla war against Allied troops. The British were forced to dispatch 12,000 troops to oppose them in the Senussi Campaign. Their rebellion was finally crushed in mid-1916.[68]

Italian participation

Further information: Battles of the Isonzo

Austro-Hungarian mountain corps in Tyrol

Italy had been allied with the German and Austro-Hungarian Empires since 1882 as part of the Triple Alliance. However, the nation had its own designs on Austrian territory in Trentino, Istria, and Dalmatia. Rome had a secret 1902 pact with France, effectively nullifying its alliance.[69] At the start of hostilities, Italy refused to commit troops, arguing that the Triple Alliance was defensive and that Austria–Hungary was an aggressor. The Austro-Hungarian government began negotiations to secure Italian neutrality, offering the French colony of Tunisia in return. The Allies made a counter-offer in which Italy would receive the Southern Tyrol, Julian March and territory on the Dalmatian coast after the defeat of Austria-Hungary. This was formalised by the Treaty of London. Further encouraged by the Allied invasion of Turkey in April 1915, Italy joined the Triple Entente and declared war on Austria-Hungary on 23 May. Fifteen months later Italy declared war on Germany.

Militarily, the Italians had numerical superiority. This advantage, however, was lost, not only because of the difficult terrain in which fighting took place, but also because of the strategies and tactics employed. Field Marshal Luigi Cadorna, a staunch proponent of the frontal assault, had dreams of breaking into the Slovenian plateau, taking Ljubljana and threatening Vienna. Cadorna's plan did not take into account the difficulties of the rugged Alpine terrain, or the technological changes that created trench warfare, giving rise to a series of bloody and inconclusive stalemated offensives.

Depiction of the Battle of Doberdò, fought in August 1916 between the Italian and Austro-Hungarian army.

On the Trentino front, the Austro-Hungarians took advantage of the mountainous terrain, which favoured the defender. After an initial strategic retreat, the front remained largely unchanged, while Austrian Kaiserschützen and Standschützen engaged Italian Alpini in bitter hand-to-hand combat throughout the summer. The Austro-Hungarians counterattacked in the Altopiano of Asiago, towards Verona and Padua, in the spring of 1916 (*Strafexpedition*), but made little progress.

Beginning in 1915, the Italians under Cadorna mounted eleven offensives on the Isonzo front along the Isonzo River, northeast of Trieste. All eleven offensives were repelled by the Austro-Hungarians, who held the higher ground. In the summer of 1916, the Italians captured the town of Gorizia. After this minor victory, the front remained static for over a year, despite several Italian offensives. In the autumn of 1917, thanks to the improving situation on the Eastern front, the Austro-Hungarian troops received large numbers of reinforcements, including German Stormtroopers and the elite Alpenkorps. The Central Powers launched a crushing offensive on 26 October 1917, spearheaded by the Germans. They achieved a victory at Caporetto. The Italian Army was routed and retreated more than 100 kilometres (**unknown operator: u'strong'** mi) to reorganise, stabilising the front at the Piave River. Since in the Battle of Caporetto the Italian Army had heavy losses, the Italian Government called to arms the so-called *'99 Boys* (*Ragazzi del '99*): that is, all males who were 18 years old. In 1918, the Austro-Hungarians failed to break through, in a series of battles on the Piave River, and were finally decisively defeated in the Battle of Vittorio Veneto in October of that year. From 5–6 November 1918, Italian forces were reported to have reached Lissa, Lagosta, Sebenico, and other localities on the Dalmatian coast.[70] By the end of hostilities in November 1918, the Italian military had seized control of the entire portion of Dalmatia that had been guaranteed to Italy by the London Pact.[71] In 1918, Admiral Enrico Millo declared himself Italy's Governor of Dalmatia.[71] Austria-Hungary surrendered in early November 1918.[72] [73]

Romanian participation

Marshal Joffre inspecting Romanian troops

Romania had been allied with the Central Powers since 1882. When the war began, however, it declared its neutrality, arguing that because Austria-Hungary had itself declared war on Serbia, Romania was under no obligation to join the war. When the Entente Powers promised Romania large territories of eastern Hungary (Transylvania and Banat) that had a large Romanian population in exchange for Romania's declaring war on the Central Powers, the Romanian government renounced its neutrality, and on 27 August 1916 the Romanian Army launched an attack against Austria-Hungary, with limited Russian support. The Romanian offensive was initially successful, pushing back the Austro-Hungarian troops in Transylvania, but a counterattack by the forces of the Central Powers drove back the Russo-Romanian forces. As a result of the Battle of Bucharest the Central Powers occupied Bucharest on 6 December 1916. Fighting in Moldova continued in 1917, resulting in a costly stalemate for the Central Powers.[74] [75] Russian withdrawal from the war in late 1917 as a result of the October Revolution meant that Romania was forced to sign an armistice with the Central Powers on 9 December 1917.

In January 1918, Romanian forces established control over Bessarabia as the Russian Army abandoned the province. Although a treaty was signed by the Romanian and the Bolshevik Russian government following talks from 5–9 March 1918 on the withdrawal of Romanian forces from Bessarabia within two months, on 27 March 1918 Romania attached Bessarabia to its territory, formally based on a resolution passed by the local assembly of the territory on the unification with Romania.

Romania officially made peace with the Central Powers by signing the Treaty of Bucharest on 7 May 1918. Under that treaty, Romania was obliged to end war with the Central Powers and make small territorial concessions to Austria-Hungary, ceding control of some passes in the Carpathian Mountains, and grant oil concessions to Germany. In exchange, the Central Powers recognised the sovereignty of Romania over Bessarabia. The treaty was renounced in October 1918 by the Alexandru Marghiloman government, and Romania nominally re-entered the war on 10 November 1918. The next day, the Treaty of Bucharest was nullified by the terms of the Armistice of Compiègne.[76] [77] Total Romanian deaths from 1914 to 1918, military and civilian, within contemporary borders, were estimated at 748,000.[78]

The role of India

Further information: Third Anglo-Afghan War and Hindu-German Conspiracy

Contrary to British fears of a revolt in India, the outbreak of the war saw an unprecedented outpouring of loyalty and good will towards the United Kingdom.[79] [80] Indian political leaders from the Indian National Congress and other groups were eager to support the British war effort since they believed that strong support for the war effort would further the cause of Indian Home Rule. The Indian Army in fact outnumbered the British Army at the beginning of the war; about 1.3 million Indian soldiers and labourers served in Europe, Africa, and the Middle East, while both the central government and the princely states sent large supplies of food, money, and ammunition. In all, 140,000 men served on the Western Front and nearly 700,000 in the Middle East. Casualties of Indian soldiers totalled 47,746 killed and 65,126 wounded during World War I.[81] The suffering engendered by the war as well as the failure of the British government to grant self-government to India after the end of hostilities bred disillusionment and fuelled the campaign for full independence that would be led by Mohandas Karamchand Gandhi and others.

Eastern Front

Russian troops awaiting a German attack

Initial actions

While the Western Front had reached stalemate, the war continued in East Europe. Initial Russian plans called for simultaneous invasions of Austrian Galicia and German East Prussia. Although Russia's initial advance into Galicia was largely successful, it was driven back from East Prussia by Hindenburg and Ludendorff at Tannenberg and the Masurian Lakes in August and September 1914.[82] [83] Russia's less developed industrial base and ineffective military leadership was instrumental in the events that unfolded. By the spring of 1915, the Russians had retreated into Galicia, and in May the Central Powers achieved a remarkable breakthrough on Poland's southern frontiers.[84] On 5 August they captured Warsaw and forced the Russians to withdraw from Poland.

Russian Revolution

Further information: North Russia Campaign

Vladimir Illyich Lenin

Despite the success of the June 1916 Brusilov Offensive in eastern Galicia,[85] dissatisfaction with the Russian government's conduct of the war grew. The offensive's success was undermined by the reluctance of other generals to commit their forces to support the victory. Allied and Russian forces were revived only temporarily by Romania's entry into the war on 27 August. German forces came to the aid of embattled Austro-Hungarian units in Transylvania, and Bucharest fell to the Central Powers on 6 December. Meanwhile, unrest grew in Russia, as the Tsar remained at the front. Empress Alexandra's increasingly incompetent rule drew protests and resulted in the murder of her favourite, Rasputin, at the end of 1916.

In March 1917, demonstrations in Petrograd culminated in the abdication of Tsar Nicholas II and the appointment of a weak Provisional Government which shared power with the Petrograd Soviet socialists. This arrangement led to confusion and chaos both at the front and at home. The army became increasingly ineffective.[84]

Russian Cossacks on the front, 1915

Discontent and the weaknesses of the Provisional Government led to a rise in popularity of the Bolshevik Party, led by Vladimir Lenin, which demanded an immediate end to the war. The successful armed uprising by the Bolsheviks of November was followed in December by an armistice and negotiations with Germany. At first the Bolsheviks refused the German terms, but when German troops began marching across the Ukraine unopposed, the new government acceded to the Treaty of Brest-Litovsk on 3 March 1918. The treaty ceded vast territories, including Finland, the Baltic provinces, parts of Poland and Ukraine to the Central Powers.[86] Despite this enormous apparent

German success, the manpower required for German occupation of former Russian territory may have contributed to the failure of the Spring Offensive and secured relatively little food or other materiel.

Signing the Treaty of Brest-Litovsk (9 February 1918) are: 1. Count Ottokar von Czernin, 2. Richard von Kühlmann, and 3. Vasil Radoslavov

With the adoption of the Treaty of Brest-Litovsk, the Entente no longer existed. The Allied powers led a small-scale invasion of Russia, partly to stop Germany from exploiting Russian resources and, to a lesser extent, to support the "Whites" (as opposed to the "Reds") in the Russian Civil War.[87] Allied troops landed in Arkhangelsk and in Vladivostok.

Central Powers proposal for starting peace negotiations

On the way to Verdun. *"They shall not pass"* is a phrase which for all time will be associated with the heroic defense of Verdun.

In December 1916, after ten brutal months of the Battle of Verdun and a successful offensive against Romania, the Germans attempted to negotiate a peace with the Allies. Soon after, U.S. President Woodrow Wilson attempted to intervene as a peacemaker, asking in a note for both sides to state their demands. Lloyd George's War Cabinet considered the German offer to be a ploy to create divisions amongst the Allies. After initial outrage and much deliberation, they took Wilson's note as a separate effort, signalling that the U.S. was on the verge of entering the war against Germany following the "submarine outrages". While the Allies debated a response to Wilson's offer, the Germans chose to rebuff it in favour of "a direct exchange of views". Learning of the German response, the Allied governments were free to make clear demands in their response of 14 January. They sought restoration of damages, the evacuation of occupied territories, reparations for France, Russia and Romania, and a recognition of the principle of nationalities. This included the liberation of Italians, Slavs, Romanians, Czecho-Slovaks, and the creation of a "free and united Poland". On the question of security, the Allies sought guarantees that would prevent or limit future wars, complete with sanctions, as a condition of any peace settlement.[88] The negotiations failed and the Entente powers rejected the German offer, because Germany did not state any specific proposals. To Wilson, the Entente powers stated that they would not start peace negotiations until the Central powers evacuated all occupied Allied territories and provided indemnities for all damage which had been done.[89]

1917–1918

French troopers under General Gouraud, with their machine guns amongst the ruins of a cathedral near the Marne, driving back the Germans. 1918

German film crew recording the action.

Developments in 1917

Events of 1917 proved decisive in ending the war, although their effects were not fully felt until 1918.

The British naval blockade began to have a serious impact on Germany. In response, in February 1917, the German General Staff convinced Chancellor Theobald von Bethmann-Hollweg to declare unrestricted submarine warfare, with the goal of starving Britain out of the war. German planners estimated that unrestricted submarine warfare would cost Britain a monthly shipping loss of 600,000 tons. The General Staff acknowledged that the policy would almost certainly bring the United States into the conflict, but calculated that British shipping losses would be so high that they would be forced to sue for peace after 5 to 6 months, before American intervention could make an impact. In reality, tonnage sunk rose above 500,000 tons per month from February to July. It peaked at 860,000 tons in April. After July, the newly re-introduced convoy system became extremely effective in reducing the U-boat threat. Britain was safe from starvation while German industrial output fell, and the United States troops joined the war in large numbers far earlier than Germany had anticipated.

On 3 May 1917, during the Nivelle Offensive, the weary French 2nd Colonial Division, veterans of the Battle of Verdun, refused their orders, arriving drunk and without their weapons. Their officers lacked the means to punish an entire division, and harsh measures were not immediately implemented. Then, mutinies afflicted an additional 54 French divisions and saw 20,000 men desert. The other Allied forces attacked but sustained tremendous casualties.[90] However, appeals to patriotism and duty, as well as mass arrests and trials, encouraged the soldiers to return to defend their trenches, although the French soldiers refused to participate in further offensive action.[91] Robert Nivelle was removed from command by 15 May, replaced by General Philippe Pétain, who suspended bloody large-scale attacks.

Haut-Rhin, France, 1917

The victory of Austria–Hungary and Germany at the Battle of Caporetto led the Allies at the Rapallo Conference to form the Supreme War Council to coordinate planning. Previously, British and French armies had operated under separate commands.

In December, the Central Powers signed an armistice with Russia. This released large numbers of German troops for use in the west. With German reinforcements and new American troops pouring in, the outcome was to be decided on the Western Front. The Central Powers knew that they could not win a protracted war, but they held high hopes for success based on a final quick offensive. Furthermore, the leaders of the Central Powers and the Allies became increasingly fearful of social unrest and revolution in Europe. Thus, both sides urgently sought a decisive victory.[92]

Entry of the United States

Non-intervention

At the outbreak of the war the United States pursued a policy of non-intervention, avoiding conflict while trying to broker a peace. When a German U-boat sank the British liner *Lusitania* on 7 May 1915 with 128 Americans among the dead, President Woodrow Wilson insisted that "America is too proud to fight" but demanded an end to attacks on passenger ships. Germany complied. Wilson unsuccessfully tried to mediate a settlement. However, he also repeatedly warned that the U.S.A. would not tolerate unrestricted submarine warfare, in violation of international law. Former president Theodore Roosevelt denounced German acts as "piracy".[93] Wilson was narrowly reelected in 1916 as his supporters emphasized "he kept us out of war".

In January 1917, Germany resumed unrestricted submarine warfare, realizing it would mean American entry. The German Foreign Minister, in the Zimmermann Telegram, invited Mexico to join the war as Germany's ally against the United States. In return, the Germans would finance Mexico's war and help it recover the territories of Texas, New Mexico, and Arizona.[94] Wilson released the Zimmerman note to the public, and Americans saw it as *casus belli*—a cause for war. Wilson called on antiwar elements to end all wars, by winning this one and eliminating militarism from the globe. He argued that the war was so important that the U.S. had to have a voice in the peace conference.[95]

President Wilson before Congress, announcing the break in official relations with Germany on 3 February 1917.

U.S. declaration of war on Germany

After the sinking of seven U.S. merchant ships by submarines and the publication of the Zimmerman telegram, Wilson called for war on Germany,[96] which the U.S. Congress declared on 6 April 1917.

First active U.S. participation

The United States was never formally a member of the Allies but became a self-styled "Associated Power". The United States had a small army, but, after the passage of the Selective Service Act, it drafted 2.8 million men,[97] and by summer 1918 was sending 10,000 fresh soldiers to France every day. In 1917, the U.S. Congress gave U.S. citizenship to Puerto Ricans when they were drafted to participate in World War I, as part of the Jones Act. Germany had miscalculated, believing it would be many more months before American soldiers would arrive and that their arrival could be stopped by U-boats.[98]

American soldiers on the Piave front hurling a shower of hand grenades into the Austrian trenches

The United States Navy sent a battleship group to Scapa Flow to join with the British Grand Fleet, destroyers to Queenstown, Ireland, and submarines to help guard convoys. Several regiments of U.S. Marines were also dispatched to France. The British and French wanted U.S. units used to reinforce their troops already on the battle lines and not waste scarce shipping on bringing over supplies. The U.S. rejected the first proposition and accepted the second. General John J. Pershing, American Expeditionary Forces (AEF) commander, refused to break up U.S. units to be used as reinforcements for British Empire and French units. As an exception, he did allow African-American combat regiments to be used in French divisions. The Harlem Hellfighters fought as part of the French 16th Division, earning a unit Croix de Guerre for their actions at Chateau-Thierry, Belleau Wood, and Sechault.[99] AEF doctrine called for the use of frontal assaults, which had long since been discarded by British Empire and French commanders because of the large loss of life.[100]

Two Allied soldiers run towards a bunker.

Austrian offer of separate peace

In 1917, Emperor Charles I of Austria secretly attempted separate peace negotiations with Clemenceau, with his wife's brother Sixtus in Belgium as an intermediary, without the knowledge of Germany. When the negotiations failed, his attempt was revealed to Germany, resulting in a diplomatic catastrophe.[101] [102]

German Spring Offensive of 1918

German General Erich Ludendorff drew up plans (codenamed Operation Michael) for the 1918 offensive on the Western Front. The Spring Offensive sought to divide the British and French forces with a series of feints and advances. The German leadership hoped to strike a decisive blow before significant U.S. forces arrived. The operation commenced on 21 March 1918 with an attack on British forces near Amiens. German forces achieved an unprecedented advance of 60 kilometres (**unknown operator: u'strong'** mi).[103]

British and French trenches were penetrated using novel infiltration tactics, also named *Hutier* tactics, after General Oskar von Hutier. Previously, attacks had been characterised by long artillery bombardments and massed assaults. However, in the Spring Offensive of 1918, Ludendorff used artillery only briefly and infiltrated small groups of infantry at weak points. They attacked command and logistics areas and bypassed points of serious resistance. More heavily armed infantry then destroyed these isolated positions. German success relied greatly on the element of surprise.[104]

The front moved to within 120 kilometres (**unknown operator: u'strong'** mi) of Paris. Three heavy Krupp railway guns fired 183 shells on the capital, causing many Parisians to flee. The initial offensive was so successful that Kaiser Wilhelm II declared 24 March a national holiday. Many Germans thought victory was near. After heavy fighting, however, the offensive was halted. Lacking tanks or motorised artillery, the Germans were unable to consolidate their gains. This situation was not helped by the supply lines now being stretched as a result of their advance.[105] The sudden stop was also a result of the four Australian Imperial Force (AIF) divisions that were "rushed" down, thus doing what no other army had done: stopping the German advance in its tracks. During that time the first Australian division was hurriedly sent north again to stop the second German breakthrough.

General Foch pressed to use the arriving American troops as individual replacements. Pershing sought instead to field American units as an independent force. These units were assigned to the depleted French and British Empire commands on 28 March. A Supreme War Council of Allied forces was created at the Doullens Conference on 5 November 1917.[106] General Foch was appointed as supreme commander of the allied forces. Haig, Petain, and Pershing retained tactical control of their respective armies; Foch assumed a coordinating rather than a directing role, and the British, French, and U.S. commands operated largely independently.[106]

British 55th (West Lancashire) Infantry Division troops blinded by tear gas during the Battle of Estaires, 10 April 1918.

Following Operation Michael, Germany launched Operation Georgette against the northern English Channel ports. The Allies halted the drive after limited territorial gains by Germany. The German Army to the south then conducted Operations Blücher and Yorck, pushing broadly towards Paris. Operation Marne was launched on 15 July, attempting to encircle Reims and beginning the Second Battle of the Marne. The resulting counterattack, starting the Hundred Days Offensive, marked the first successful Allied offensive of the war.

By 20 July the Germans were back across the Marne at their Kaiserschlacht starting lines,[107] having achieved nothing. Following this last phase of the war in the West, the German Army never regained the initiative. German casualties between March and April 1918 were 270,000, including many highly trained storm troopers.

Meanwhile, Germany was falling apart at home. Anti-war marches became frequent and morale in the army fell. Industrial output was 53 percent of 1913 levels.

New states under war zone

In the late spring of 1918, three new states were formed in the South Caucasus: the Democratic Republic of Armenia, the Azerbaijan Democratic Republic, and the Democratic Republic of Georgia, which declared their independence from the Russian Empire.[108] Two other minor entities were established, the Centrocaspian Dictatorship and South West Caucasian Republic (the former was liquidated by Azerbaijan in the autumn of 1918 and the latter by a joint Armenian-British task force in early 1919). With the withdrawal of the Russian armies from the Caucasus front in the winter of 1917–18, the three major republics braced for an imminent Ottoman advance, which commenced in the early months of 1918. Solidarity was briefly maintained when the Transcaucasian Federative Republic was created in the spring of 1918 but collapsed in May, when the Georgians asked and received protection from Germany and the Azerbaijanis concluded a treaty with the Ottoman Empire that was more akin to a military alliance. Armenia was left to fend for itself and struggled for five months against the threat of a full-fledged occupation by the Ottoman Turks.[109]

Allied victory: summer and autumn 1918

The Allied counteroffensive, known as the Hundred Days Offensive, began on 8 August 1918. The Battle of Amiens developed with III Corps British Fourth Army on the left, the French First Army on the right, and the Australian and Canadian Corps spearheading the offensive in the centre through Harbonnières.[110] [111] It involved 414 tanks of the Mark IV and Mark V type, and 120,000 men. They advanced 12 kilometres (**unknown operator: u'strong'** mi) into German-held territory in just seven hours. Erich Ludendorff referred to this day as the "Black Day of the German army".[110] [112]

Aerial view of ruins of Vaux, France, 1918

The Australian-Canadian spearhead at Amiens, a battle that was the beginning of Germany's downfall,[42] helped pull forward the British armies to the north and the French armies to the south. On the British Fourth Army front at Amiens, after an advance as far as 14 miles (**unknown operator: u'strong'** km), German resistance stiffened, and the battle there concluded. But the French Third Army lengthened the Amiens front on 10 August, when it was thrown in on the right of the French First Army, and advanced 4 miles (**unknown operator: u'strong'** km), liberating Lassigny in fighting which lasted until 16 August. South of the French Third Army, General Charles Mangin (The Butcher) drove his French Tenth Army forward at Soissons on 20 August to capture eight thousand prisoners, two hundred guns, and the Aisne heights overlooking and menacing the German position north of the Vesle.[113] Another "Black day", as described by Erich Ludendorff.

Meanwhile General Byng of the British Third Army, reporting that the enemy on his front was thinning in a limited withdrawal, was ordered to attack with 200 tanks towards Bapaume, opening the Battle of Albert, with specific orders "To break the enemy's front, in order to outflank the enemy's present battle front" (opposite the British Fourth Army at Amiens).[42] Allied leaders had now realised that to continue an attack after resistance had hardened was a waste of lives, and it was better to turn a line than to try to roll over it. They began to undertake attacks in quick order to take advantage of successful advances on the flanks, then broke them off when each attack lost its initial impetus.[113]

The British Third Army's 15-mile (**unknown operator: u'strong'** km) front north of Albert progressed after stalling for a day against the main resistance line to which the enemy had withdrawn.[114] Rawlinson's British Fourth Army was able to push its left flank forward between Albert and the Somme, straightening the line between the advanced positions of the Third Army and the Amiens front, which resulted in recapturing Albert at the same time.[113] On 26 August the British First Army on the left of the Third Army was drawn into the battle, extending it northward to beyond Arras. The Canadian Corps, already back in the vanguard of the First Army, fought its way from Arras eastward 5 miles (**unknown operator: u'strong'** km) astride the heavily defended Arras-Cambrai area before reaching the outer defences of the Hindenburg Line, breaching them on the 28 and 29 August. Bapaume fell on 29 August to the New Zealand Division of the Third Army, and the Australians, still leading the advance of the Fourth Army, were again able to push forward at Amiens to take Peronne and Mont Saint-Quentin on 31 August. Further south, the French First and Third Armies had slowly fought forward while the Tenth Army, which had by now crossed the Ailette and was east of the Chemin des Dames, neared the Alberich position of the Hindenburg Line.[115] During the last week of August the pressure along a 70-mile (**unknown operator: u'strong'** km) front against the enemy was heavy and unrelenting. From German accounts, "Each day was spent in bloody fighting against an ever and again on-storming enemy, and nights passed without sleep in retirements to new lines."[113] Even to the north in Flanders the British Second and Fifth Armies during August and September were able to make progress, taking prisoners and positions that had previously been denied them.[115]

On 2 September the Canadian Corps outflanking of the Hindenburg line, with the breaching of the Wotan Position, made it possible for the Third Army to advance, which sent repercussions all along the Western Front. That same day Oberste Heeresleitung (OHL) had no choice but to issue orders to six armies to withdraw back into the Hindenburg Line in the south, behind the Canal du Nord on the Canadian-First Army's front and back to a line east of the Lys in the north. This ceded without a fight the salient seized the previous April.[116] According to Ludendorff "We had to admit the necessity ...to withdraw the entire front from the Scarpe to the Vesle."[117]

American troops in Vladivostok, Siberia, August 1918

In nearly four weeks of fighting beginning 8 August, over 100,000 German prisoners were taken, 75,000 by the BEF and the rest by the French. As of "The Black Day of the German Army", the German High Command realised the war was lost and made attempts to reach a satisfactory end. The day after that battle Ludenforff told Colonel Mertz: "We cannot win the war any more, but we must not lose it either." On 11 August he offered his resignation to the Kaiser, who refused it, replying, "I see that we must strike a balance. We have nearly reached the limit of our powers of resistance. The war must be ended." On 13 August at Spa, Hindenburg, Ludendorff, the Chancellor, and Foreign Minister Hintz agreed that the war could not be ended militarily, and on the following day the German Crown Council decided that victory in the field was now most improbable. Austria and Hungary warned that they could only continue the war until December, and Ludendorff recommended immediate peace negotiations, to which the Kaiser responded by instructing Hintz to seek the mediation of the Queen of the Netherlands. Prince Rupprecht warned Prince Max of Baden: "Our military situation has deteriorated so rapidly that I no longer believe we can hold out over the winter; it is even possible that a catastrophe will come earlier." On 10 September Hindenburg urged peace moves to Emperor Charles of Austria, and Germany appealed to the Netherlands for mediation. On 14 September Austria sent a note to all belligerents and neutrals suggesting a meeting for peace talks on neutral soil, and on 15 September Germany made a peace offer to Belgium. Both peace offers were rejected, and on 24 September OHL informed the leaders in Berlin that armistice talks were inevitable.[115]

Close-up view of an American major in the basket of an observation balloon flying over territory near front lines

September saw the Germans continuing to fight strong rear-guard actions and launching numerous counterattacks on lost positions, but only a few succeeded, and then only temporarily. Contested towns, villages, heights, and trenches in the screening positions and outposts of the Hindenburg Line continued to fall to the Allies, with the BEF alone taking 30,441 prisoners in the last week of September. Further small advances eastward would follow the Third Army's victory at Ivincourt on 12 September, the Fourth Army's at Epheny on 18 September, and the French gain of Essigny-le-Grand a day later. On 24 September a final assault by both the British and French on a 4-mile (**unknown operator: u'strong'** km) front would come within 2 miles (**unknown operator: u'strong'** km) of St. Quentin.[115] With the outposts and preliminary defensive lines of the Siegfried and Alberich Positions eliminated, the Germans were now completely back in the Hindenburg Line. With the Wotan position of that line already breached and the Siegfried position in danger of being turned from the north, the time had now come for an Allied assault on the whole length of the line.

The Allied attack on the Hindenburg Line, begun on 26 September, included U.S. soldiers. The still-green American troops suffered problems coping with supply trains for large units on a difficult landscape.[118] The following week cooperating French and American units broke through in Champagne at the Battle of Blanc Mont Ridge, forcing the Germans off the commanding heights, and closing towards the Belgian frontier.[119] The last Belgian town to be liberated before the armistice was Ghent, which the Germans held as a pivot until the Allies brought up artillery.[120][121] The German army had to shorten its front and use the Dutch frontier as an anchor to fight rear-guard actions.

Men of U.S. 64th Regiment, 7th Infantry Division, celebrate the news of the Armistice, 11 November 1918

When Bulgaria signed a separate armistice on 29 September, the Allies gained control of Serbia and Greece. Ludendorff, having been under great stress for months, suffered something similar to a breakdown. It was evident that Germany could no longer mount a successful defence.[122] [123]

Meanwhile, news of Germany's impending military defeat spread throughout the German armed forces. The threat of mutiny was rife. Admiral Reinhard Scheer and Ludendorff decided to launch a last attempt to restore the "valour" of the German Navy. Knowing the government of Prince Maximilian of Baden would veto any such action, Ludendorff decided not to inform him. Nonetheless, word of the impending assault reached sailors at Kiel. Many, refusing to be part of a naval offensive which they believed to be suicidal, rebelled and were arrested. Ludendorff took the blame; the Kaiser dismissed him on 26 October. The collapse of the Balkans meant that Germany was about to lose its main supplies of oil and food. Its reserves had been used up, even as U.S. troops kept arriving at the rate of 10,000 per day.[124]

Having suffered over 6 million casualties, Germany moved towards peace. Prince Maximilian of Baden took charge of a new government as Chancellor of Germany to negotiate with the Allies. Telegraphic negotiations with President Wilson began immediately, in the vain hope that he would offer better terms than the British and French. Instead Wilson demanded the abdication of the Kaiser. There was no resistance when the Social Democrat Philipp Scheidemann on 9 November declared Germany to be a republic. Imperial Germany was dead; a new Germany had been born: the Weimar Republic.[125]

Armistices and capitulations

The signing of the armistice.

The collapse of the Central Powers came swiftly. Bulgaria was the first to sign an armistice, on 29 September 1918 at Saloniki.[127] On 30 October, the Ottoman Empire capitulated at Moudros (Armistice of Mudros).[127]

On 24 October, the Italians began a push which rapidly recovered territory lost after the Battle of Caporetto. This culminated in the Battle of Vittorio Veneto, which marked the end of the Austro-Hungarian Army as an effective fighting force. The offensive also triggered the disintegration of the Austro-Hungarian Empire. During the last week of October, declarations of independence were made in Budapest, Prague, and Zagreb. On 29 October, the imperial authorities asked Italy for an armistice. But the Italians continued advancing, reaching Trento, Udine, and Trieste. On 3 November Austria–Hungary sent a flag of truce to ask for an Armistice. The terms, arranged by telegraph with the Allied Authorities in Paris, were communicated to the Austrian commander and accepted. The Armistice with Austria was signed in the Villa Giusti, near Padua, on 3 November. Austria and Hungary signed separate armistices following the overthrow of the Habsburg Monarchy.

Following the outbreak of the German Revolution of 1918–1919, a republic was proclaimed on 9 November. The Kaiser fled to the Netherlands. On 11 November an armistice with Germany was signed in a railroad carriage at Compiègne. At 11 am on 11 November 1918—"the eleventh hour of the eleventh day of the eleventh month"—a ceasefire came into effect. Opposing armies on the Western Front began to withdraw from their positions. Canadian Private George Lawrence Price was shot by a German sniper at 10:57 and died at 10:58.[128] American Henry Gunther was killed 60 seconds before the armistice came into force while charging astonished German troops who were aware the Armistice was nearly upon them.[129] The last British soldier to die was Pte George Edwin Ellison. The last casualty of the war was a German, Lieutenant Thomas, who, after 11 am, was walking towards the line to inform Americans who had not yet been informed of the Armistice that they would be vacating the buildings behind them.[130] The occupation of the Rhineland took place following the Armistice. The occupying armies consisted of American, Belgian, British and French forces.

In the forest of Compiègne after agreeing to the armistice that ended the war, Foch is seen second from the right. The carriage seen in the background, where the armistice was signed, was later chosen as the symbolic setting of Pétain's June 1940 armistice. It was moved to Berlin as a prize, but because of Allied bombing was eventually moved to Crawinkel, Thuringia, where it was deliberately destroyed by SS troops in 1945.[126]

Allied superiority and the stab-in-the-back legend, November 1918

In November 1918 the Allies had ample supplies of men and materiel to invade Germany. Yet at the time of the armistice, no Allied force had crossed the German frontier; the Western Front was still almost 900 mi (**unknown operator: u'strong'** km) from Berlin; and the Kaiser's armies had retreated from the battlefield in good order. These factors enabled Hindenburg and other senior German leaders to spread the story that their armies had not really been defeated. This resulted in the stab-in-the-back legend,[131] [132] which attributed Germany's defeat not to its inability to continue fighting (even though up to a million soldiers were suffering from the 1918 flu pandemic and unfit to fight), but to the public's failure to respond to its "patriotic calling" and the supposed intentional sabotage of the war effort, particularly by Jews, Socialists, and Bolsheviks.

A formal state of war between the two sides persisted for another seven months, until the signing of the Treaty of Versailles with Germany on 28 June 1919. Later treaties with Austria, Hungary, Bulgaria, and the Ottoman Empire were signed. However, the negotiation of the latter treaty with the Ottoman Empire was followed by strife (the Turkish War of Independence), and a final peace treaty between the Allied Powers and the country that would shortly become the Republic of Turkey was not signed until 24 July 1923, at Lausanne.

Some war memorials date the end of the war as being when the Versailles Treaty was signed in 1919, which was when many of the troops serving abroad finally returned to their home countries; by contrast, most commemorations of the war's end concentrate on the armistice of 11 November 1918. Legally, the formal peace treaties were not complete until the last, the Treaty of Lausanne, was signed. Under its terms, the Allied forces divested Constantinople on 23 August 1923.

Technology

Armoured cars

The First World War began as a clash of 20th-century technology and 19th-century tactics, with the inevitably large ensuing casualties. By the end of 1917, however, the major armies, now numbering millions of men, had modernised and were making use of telephone, wireless communication,[133] armoured cars, tanks,[134] and aircraft. Infantry formations were reorganised, so that 100-man companies were no longer the main unit of manoeuvre; instead, squads of 10 or so men, under the command of a junior NCO, were favoured.

Artillery also underwent a revolution. In 1914, cannons were positioned in the front line and fired directly at their targets. By 1917, indirect fire with guns (as well as mortars and even machine guns) was commonplace, using new techniques for spotting and ranging, notably aircraft and the often overlooked field telephone. Counter-battery missions became commonplace, also, and sound detection was used to locate enemy batteries.

Germany was far ahead of the Allies in utilising heavy indirect fire. The German Army employed 150 and 210 mm howitzers in 1914, when typical French and British guns were only 75 and 105 mm. The British had a 6 inch (152 mm) howitzer, but it was so heavy it had to be hauled to the field in pieces and assembled. Germans also fielded Austrian 305 mm and 420 mm guns, and already by the beginning of the war had inventories of various calibers of *Minenwerfer* ideally suited for trench warfare.[135]

Much of the combat involved trench warfare, in which hundreds often died for each yard gained. Many of the deadliest battles in history occurred during the First World War. Such battles include Ypres, the Marne, Cambrai, the Somme, Verdun, and Gallipoli. The Germans employed the Haber process of nitrogen fixation to provide their forces with a constant supply of gunpowder, despite the British naval blockade.[136] Artillery was responsible for the largest number of casualties[137] and consumed vast quantities of explosives. The large number of head wounds caused by exploding shells and fragmentation forced the combatant nations to develop the modern steel helmet, led by the French, who introduced the Adrian helmet in 1915. It was quickly followed by the Brodie helmet, worn by British Imperial and U.S. troops, and in 1916 by the distinctive German *Stahlhelm*, a design, with improvements, still in use today.

> *"Gas! Gas! Quick, boys!... Fitting the clumsy helmets just in time; But someone still was yelling out and stumbling, And flound'ring like a man in fire or lime... Dim, through the misty panes and thick green light, As under a green sea, I saw him drowning."*- Wilfred Owen, *DULCE ET DECORUM EST*, 1917[138]

The widespread use of chemical warfare was a distinguishing feature of the conflict. Gases used included chlorine, mustard gas and phosgene. Few war casualties were caused by gas,[139] as effective countermeasures to gas attacks were quickly created, such as gas masks. The use of chemical warfare and small-scale strategic bombing were both outlawed by the 1907 Hague Conventions, and both proved to be of limited effectiveness,[140] though they captured the public imagination.[141]

The most powerful land-based weapons were railway guns weighing hundreds of tons apiece. These were nicknamed Big Berthas, even though the namesake was not a railway gun. Germany developed the Paris Gun, able to bombard Paris from over 100 kilometres (**unknown operator: u'strong'** mi), though shells were relatively light at 94 kilograms (210 lb). While the Allies also had railway guns, German models severely out-ranged and out-classed them.

Aviation

Fixed-wing aircraft were first used militarily by the Italians in Libya on 23 October 1911 during the Italo-Turkish War for reconnaissance, soon followed by the dropping of grenades and aerial photography the next year. By 1914 their military utility was obvious. They were initially used for reconnaissance and ground attack. To shoot down enemy planes, anti-aircraft guns and fighter aircraft were developed. Strategic bombers were created, principally by the Germans and British, though the former used Zeppelins as well.[143] Towards the end of the conflict, aircraft carriers were used for the first time, with HMS *Furious* launching Sopwith Camels in a raid to destroy the Zeppelin hangars at Tondern in 1918.[144]

RAF Sopwith Camel. In April 1917, the average life expectancy of a British pilot on the Western Front was 93 flying hours.[142]

Johnson's Nieuport 16 armed with Le Prieur rockets for attacking observation balloons.

Manned observation balloons, floating high above the trenches, were used as stationary reconnaissance platforms, reporting enemy movements and directing artillery. Balloons commonly had a crew of two, equipped with parachutes.,[145] so that if there was an enemy air attack the crew could parachute to safety. (At the time, parachutes were too heavy to be used by pilots of aircraft (with their marginal power output), and smaller versions were not developed until the end of the war; they were also opposed by British leadership, who feared they might promote cowardice.)[146]

Recognised for their value as observation platforms, balloons were important targets of enemy aircraft. To defend them against air attack, they were heavily protected by antiaircraft guns and patrolled by friendly aircraft; to attack them, unusual weapons such as air-to-air rockets were even tried. Thus, the reconnaissance value of blimps and balloons contributed to the development of air-to-air combat between all types of aircraft, and to the trench stalemate, because it was impossible to move large numbers of troops undetected. The Germans conducted air raids on England during 1915 and 1916 with airships, hoping to damage British morale and cause aircraft to be diverted from the front lines, and indeed the resulting panic led to the diversion of several squadrons of fighters from France.[143] [146]

German trench destroyed by a mine explosion. Approximately 10,000 German troops were killed when the 19 mines were simultaneously detonated.

Improvements in naval technology during World War I

Germany deployed U-boats (submarines) after the war began. Alternating between restricted and unrestricted submarine warfare in the Atlantic, the Kaiserliche Marine employed them to deprive the British Isles of vital supplies. The deaths of British merchant sailors and the seeming invulnerability of U-boats led to the development of depth charges (1916), hydrophones (passive sonar, 1917), blimps, hunter-killer submarines (HMS *R-1*, 1917), forward-throwing anti-submarine weapons, and dipping hydrophones (the latter two both abandoned in 1918).[147] To extend their operations, the Germans proposed supply submarines (1916). Most of these would be forgotten in the interwar period until World War II revived the need.

Improvements in ground warfare technology in World War I

British Vickers machine gun

Trenches, machine guns, air reconnaissance, barbed wire, and modern artillery with fragmentation shells helped bring the battle lines of World War I to a stalemate. The British sought a solution with the creation of the tank and mechanised warfare. The first tanks were used during the Battle of the Somme on 15 September 1916. Mechanical reliability was an issue, but the experiment proved its worth. Within a year, the British were fielding tanks by the hundreds, and they showed their potential during the Battle of Cambrai in November 1917, by breaking the Hindenburg Line, while combined arms teams captured 8000 enemy soldiers and 100 guns. The conflict also saw the introduction of Light automatic weapons and submachine guns, such as the Lewis Gun, the Browning automatic rifle, and the Bergmann MP18.

Flamethrowers and subterranean transport

Another new weapon, the flamethrower, was first used by the German army and later adopted by other forces. Although not of high tactical value, the flamethrower was a powerful, demoralising weapon that caused terror on the battlefield. It was a dangerous weapon to wield, as its heavy weight made operators vulnerable targets.

Trench railways evolved to supply the enormous quantities of food, water, and ammunition required to support large numbers of soldiers in areas where conventional transportation systems had been destroyed. Internal combustion engines and improved traction systems for automobiles and trucks/lorries eventually rendered trench railways obsolete.

War crimes

Genocide and ethnic cleansing

Austro-Hungarian soldiers executing Serb civilians during World War I occupation, Mačva, 1914

The ethnic cleansing of the Ottoman Empire's Armenian population, including mass deportations and executions, during the final years of the Ottoman Empire is considered genocide.[148] The Ottomans saw the entire Armenian population as an enemy[149] that had chosen to side with Russia at the beginning of the war.[150] In early 1915, a number of Armenians joined the Russian forces, and the Ottoman government used this as a pretext to issue the Tehcir Law (Law on Deportation). This authorized the deportation of the Armenians from eastern provinces of the Empire to Syria between 1915 and 1917. The exact number of deaths is unknown: while Balakian gives a range of 250,000 to 1.5 million for the deaths of Armenians,[151] the International Association of Genocide Scholars estimates over 1 million.[148] The government of Turkey has consistently rejected charges of genocide, arguing that those who died were victims of inter-ethnic fighting, famine, or disease during the First World War.[152] Other ethnic groups were similarly attacked by the Ottoman Empire during this period, including Assyrians and Greeks, and some scholars consider those events to be part of the same policy of extermination.[153] [154] [155]

Russian Empire

Remains of Armenians burnt alive in the cattle shed in Aly-Zrna, 1915

Approximately 200,000 Germans living in Volhynia and about 600,000 Jews were deported by the Russian authorities.[156] [157] [158] In 1916, an order was issued to deport around 650,000 Volga Germans to the east as well, but the Russian Revolution prevented this from being carried out.[159] Many pogroms accompanied the Revolution of 1917 and the ensuing Russian Civil War, 60,000–200,000 civilian Jews were killed in the atrocities throughout the former Russian Empire.[160] [161]

"Rape of Belgium"

The German invaders treated any resistance—such as sabotaging rail lines—as illegal and immoral, and shot the offenders and burned buildings in retaliation. The German army executed over 6,500 French and Belgian civilians between August and November 1914, usually in near-random large-scale shootings of civilians ordered by junior German officers. The German Army destroyed 15,000-20,000 buildings—most famously the university library at Louvain—and generated a refugee wave of over a million people. Over half the German regiments in Belgium were involved in major incidents.[162] Thousands of workers were shipped to Germany to work in factories. British propaganda dramatizing the "Rape of Belgium" attracted much attention in the U.S., while Berlin said it was legal and necessary because of the threat of "franc-tireurs" (guerrillas) like those in France in 1870.[163] The British and French magnified the reports and disseminated them at home and in the U.S., where they played a major role in dissolving support for Germany.[164] [165]

Soldiers' experiences

The First Contingent of the Bermuda Volunteer Rifle Corps to the 1 Lincolns, training in Bermuda for the Western Front, winter 1914–1915. One in four survived the war.

The soldiers of the war were initially volunteers, except for those of Italy, but increasingly were conscripted into service. Britain's Imperial War Museum has collected more than 2,500 recordings of soldiers' personal accounts, and selected transcripts, edited by military author Max Arthur, have been published. The Museum believes that historians have not taken full account of this material, and accordingly has made the full archive of recordings available to authors and researchers.[166] Surviving veterans, returning home, often found that they could only discuss their experiences amongst themselves. Grouping together, they formed "veterans' associations" or "Legions".

Prisoners of war

This photograph shows an emaciated Indian Army soldier who survived the Siege of Kut.

About 8 million men surrendered and were held in POW camps during the war. All nations pledged to follow the Hague Conventions on fair treatment of prisoners of war. POWs' rate of survival was generally much higher than that of their peers at the front.[167] Individual surrenders were uncommon; large units usually surrendered en masse. At the Battle of Tannenberg 92,000 Russians surrendered. When the besieged garrison of Kaunas surrendered in 1915, some 20,000 Russians became prisoners. Over half of Russian losses (as a proportion of those captured, wounded, or killed) were to prisoner status; for Austria-Hungary 32%, for Italy 26%, for France 12%, for Germany 9%; for Britain 7%. Prisoners from the Allied armies totalled about 1.4 million (not including Russia, which lost 2.-3.5 million men as prisoners.) From the Central Powers about 3.3 million men became prisoners.[168]

Germany held 2.5 million prisoners; Russia held 2.9 million; while Britain and France held about 720,000. Most were captured just prior to the Armistice. The U.S. held 48,000. The most dangerous moment was the act of surrender, when helpless soldiers were sometimes gunned down.[169] [170] Once prisoners reached a camp, conditions were, in general, satisfactory (and much better than in World War II), thanks in part to the efforts of the International Red Cross and inspections by neutral nations. However, conditions were terrible in Russia: starvation was common for prisoners and civilians alike; about 15–20% of the prisoners in Russia died. In Germany, food was scarce, but only 5% died.[171] [172] [173]

German prisoners in a French prison camp

The Ottoman Empire often treated POWs poorly.[174] Some 11,800 British Empire soldiers, most of them Indians, became prisoners after the Siege of Kut in Mesopotamia in April 1916; 4,250 died in captivity.[175] Although many were in very bad condition when captured, Ottoman officers forced them to march 1100 kilometres (**unknown operator: u'strong'** mi) to Anatolia. A survivor said: "We were driven along like beasts; to drop out was to die."[176] The survivors were then forced to build a railway through the Taurus Mountains.

In Russia, when the prisoners from the Czech Legion of the Austro-Hungarian army were released in 1917, they re-armed themselves and briefly became a military and diplomatic force during the Russian Civil War.

While the Allied prisoners of the Central Powers were quickly sent home at the end of active hostilities, the same treatment was not granted to Central Power prisoners of the Allies and Russia, many of whom served as forced labor, e.g., in France until 1920. They were released only after many approaches by the Red Cross to the Allied Supreme Council.[177] German prisoners were still being held in Russia as late as 1924.[178]

Military attachés and war correspondents

Military and civilian observers from every major power closely followed the course of the war. Many were able to report on events from a perspective somewhat akin to modern "embedded" positions within the opposing land and naval forces. These military attachés and other observers prepared voluminous first-hand accounts of the war and analytical papers.

For example, former U.S. Army Captain Granville Fortescue followed the developments of the Gallipoli Campaign from an embedded perspective within the ranks of the Turkish defenders; and his report was passed through Turkish censors before being printed in London and New York.[179] However, this observer's role was abandoned when the U.S. entered the war, as Fortescue immediately re-enlisted, sustaining wounds at Forest of Argonne in the Meuse-Argonne Offensive, September 1918.[180]

In-depth observer narratives of the war and more narrowly focused professional journal articles were written soon after the war; and these post-war reports conclusively illustrated the battlefield destructiveness of this conflict. This was not the first time the tactics of entrenched positions for infantry defended with machine guns and artillery became vitally important. The Russo-Japanese War had been closely observed by military attachés, war correspondents and other observers; but, from a 21st century perspective, it is now apparent that a range of tactical lessons were disregarded or not used in the preparations for war in Europe and throughout the Great War.[181]

Support and opposition to the war

Support

In the Balkans, Yugoslav nationalists such as the leader Ante Trumbić in the Balkans strongly supported the war, desiring the freedom of Yugoslavs from Austria-Hungary and other foreign powers and the creation of an independent Yugoslavia.[182] The Yugoslav Committee was formed in Paris on 30 April 1915 but shortly moved its office to London; Trumbić led the Committee.[182]

Old England first, self second 1916

In the Middle East, Arab nationalism soared in Ottoman territories in response to the rise of Turkish nationalism during the war, with Arab nationalist leaders advocating the creation of a pan-Arab state.[183] In 1916, the Arab Revolt began in Ottoman-controlled territories of the Middle East in an effort to achieve independence.[183]

Italian nationalism was stirred by the outbreak of the war and was initially strongly supported by a variety of political factions. One of the most prominent and popular Italian nationalist supporters of the war was Gabriele d'Annunzio, who promoted Italian irredentism and helped sway the Italian public to support intervention in the war.[184] The Italian Liberal Party under the leadership of Paolo Boselli promoted intervention in the war on the side of the Allies and utilised the Dante Aligheri Society to promote Italian nationalism.[185]

A number of socialist parties initially supported the war when it began in August 1914.[186] But European socialists split on national lines, with the concept of class conflict held by radical socialists such as Marxists and syndicalists being overborne by their patriotic support for war.[187] Once the war began, Austrian, British, French, German, and Russian socialists followed the rising nationalist current by supporting their countries' intervention in the war.[188]

Italian socialists were divided on whether to support the war or oppose it; some were militant supporters of the war, including Benito Mussolini and Leonida Bissolati.[189] However, the Italian Socialist Party decided to oppose the war after anti-militarist protestors were killed, resulting in a general strike called Red Week.[190] The Italian Socialist

Party purged itself of pro-war nationalist members, including Mussolini.[190] Mussolini, a syndicalist who supported the war on grounds of irredentist claims on Italian-populated regions of Austria-Hungary, formed the pro-interventionist *Il Popolo d'Italia* and the *Fasci Riviluzionario d'Azione Internazionalista* ("Revolutionary Fasci for International Action") in October 1914 that later developed into the *Fasci di Combattimento* in 1919, the origin of fascism.[191] Mussolini's nationalism enabled him to raise funds from Ansaldo (an armaments firm) and other companies to create *Il Popolo d'Italia* to convince socialists and revolutionaries to support the war.[192]

In April 1918 the Rome Congress of Oppressed Nationalities met, including Czechoslovak, Italian, Polish, Transylvanian, and Yugoslav representatives who urged the Allies to support national self-determination for the peoples residing within Austria-Hungary.[186]

Opposition

The trade union and socialist movements had long voiced their opposition to a war, which they argued would mean only that workers would kill other workers in the interest of capitalism. Once war was declared, however, many socialists and trade unions backed their governments. Among the exceptions were the Bolsheviks, the Socialist Party of America, and the Italian Socialist Party, and individuals such as Karl Liebknecht, Rosa Luxemburg, and their followers in Germany. There were also small anti-war groups in Britain and France.

Shortly before the war, British General Horace Smith-Dorrien predicted a catastrophic war which should be avoided at *almost any cost.*

Benedict XV, elected to the papacy less than three months into World War I, made the war and its consequences the main focus of his early pontificate. In stark contrast to his predecessor,[193] five days after his election he spoke of his determination to do what he could to bring peace. His first encyclical, Ad Beatissimi Apostolorum, given 1 November 1914, was concerned with this subject. Seen as being biased in favour of the other and resented for weakening national morale, Benedict XV found his abilities and unique position as a religious emissary of peace ignored by the belligerent powers.

The 1915 Treaty of London between Italy and the Triple Entente included secret provisions whereby the Allies agreed with Italy to ignore papal peace moves towards the Central Powers. Consequently, the publication of Benedict's proposed seven-point Peace Note of August 1917 was roundly ignored by all parties except Austria-Hungary.[194]

The Deserter, 1916. Anti-war cartoon depicting Jesus facing a firing squad made up of soldiers from five different European countries.

In Britain, in 1914, the Public Schools Officers' Training Corps annual camp was held at Tidworth Pennings, near Salisbury Plain. Head of the British Army Lord Kitchener was to review the cadets, but the imminence of the war prevented him. General Horace Smith-Dorrien was sent instead. He surprised the two-or-three thousand cadets by declaring (in the words of Donald Christopher Smith, a Bermudian cadet who was present), *that war should be avoided at almost any cost, that war would solve nothing, that the whole of Europe and more besides would be reduced to ruin, and that the loss of life would be so large that whole populations would be decimated. In our ignorance I, and many of us, felt almost ashamed of a British General who uttered such depressing and unpatriotic sentiments, but during the next four years, those of us who survived the holocaust—probably not more than one-quarter of us—learned how right the General's prognosis was and how courageous he had been to utter it.*[195] Voicing these sentiments did not hinder Smith-Dorien's career, or prevent him from doing his duty in World War I to the best of his abilities.

1917 – Execution at Verdun at the time of the mutinies.

Many countries jailed those who spoke out against the conflict. These included Eugene Debs in the United States and Bertrand Russell in Britain. In the U.S., the Espionage Act of 1917 and Sedition Act of 1918 made it a federal crime to oppose military recruitment or make any statements deemed "disloyal". Publications at all critical of the government were removed from circulation by postal censors,[95] and many served long prison sentences for statements of fact deemed unpatriotic.

A number of nationalists opposed intervention, particularly within states that the nationalists were hostile to. Irish nationalists staunchly opposed taking part in the intervention of the United Kingdom of Great Britain and Ireland.[196] The war began amid the Home Rule crisis in Ireland that had begun in 1912, and by 1914 there was a serious possibility of an outbreak of civil war in Ireland between Irish unionists and republicans.[196] Irish nationalists and Marxists attempted to pursue Irish independence, culminating in the Easter Rising of 1916, with Germany sending 20,000 rifles to Ireland in order to stir unrest in the United Kingdom.[196] The UK government placed Ireland under martial law in response to the Easter Rising.[197]

The revolt of Czech units in Rumburk in May 1918 was brutally suppressed, and its leaders executed.

Other opposition came from conscientious objectors – some socialist, some religious – who refused to fight. In Britain 16,000 people asked for conscientious objector status.[198] Many suffered years of prison, including solitary confinement and bread and water diets. Even after the war, in Britain many job advertisements were marked "No conscientious objectors need apply".

The Central Asian Revolt started in the summer of 1916, when the Russian Empire government ended its exemption of Muslims from military service.[199]

In 1917, a series of mutinies in the French army led to dozens of soldiers being executed and many more imprisoned.

In Milan in May 1917, Bolshevik revolutionaries organised and engaged in rioting calling for an end to the war, and managed to close down factories and stop public transportation.[200] The Italian army was forced to enter Milan with tanks and machine guns to face Bolsheviks and anarchists, who fought violently until 23 May when the army gained control of the city. Almost fifty people (including three Italian soldiers) were killed and over 800 people arrested.[200]

The Conscription Crisis of 1917 in Canada erupted when conservative Prime Minister Robert Borden brought in compulsory military service over the objection of French-speaking Quebecers.[201] Out of approximately 625,000 Canadians who served, about 60,000 were killed and another 173,000 wounded.[202]

German Revolution, November 1918

In 1917, Emperor Charles I of Austria secretly entered into peace negotiations with the Allied powers, with his brother-in-law Sixtus as intermediary, without the knowledge of his ally Germany. He failed, however, because of the resistance of Italy.[203]

In September 1917, Russian soldiers in France began questioning why they were fighting for the French at all and mutinied.[204] In Russia, opposition to the war led to soldiers also establishing their own revolutionary committees, which helped foment the October Revolution of 1917, with the call going up for "bread, land, and peace". The Bolsheviks agreed to a peace treaty with Germany, the peace of Brest-Litovsk, despite its harsh conditions.

In northern Germany, the end of October 1918, saw the beginning of the German Revolution of 1918–1919. Units of the German Navy refused to set sail for a last, large-scale operation in a war which they saw as good as lost; this initiated the uprising. The sailors' revolt which then ensued in the naval ports of Wilhelmshaven and Kiel spread across the whole country within days and led to the proclamation of a republic on 9 November 1918 and shortly thereafter to the abdication of Kaiser Wilhelm II.

Conscription

As the war slowly turned into a war of attrition, conscription was implemented in some countries. This issue was particularly explosive in Canada and Australia. In the former it opened a political gap between French-Canadians, who claimed their true loyalty was to Canada and not the British Empire, and members of the Anglophone majority, who saw the war as a duty to both Britain and Canada. Prime Minister Robert Borden pushed through a Military Service Act, provoking the Conscription Crisis of 1917. In Australia, a sustained pro-conscription campaign by Prime Minister Billy Hughes caused a split in the Australian Labor Party, so Hughes formed the Nationalist Party of Australia in 1917 to pursue the matter. Nevertheless, the labour movement, the Catholic Church, and Irish nationalist expatriates successfully opposed Hughes' push, which was rejected in two plebiscites.

Conscription put into uniform nearly every physically fit man in Britain, six of ten million eligible. Of these, about 750,000 lost their lives and 1,700,000 were wounded. Most deaths were to young unmarried men; however, 160,000 wives lost husbands and 300,000 children lost fathers.[205]

Aftermath

Health and economic effects

No other war had changed the map of Europe so dramatically. Four empires disappeared: the German, Austro-Hungarian, Ottoman, and Russian. Four dynasties, together with their ancillary aristocracies, all fell after the war: the Hohenzollerns, the Habsburgs, the Romanovs, and the Ottomans. Belgium and Serbia were badly damaged, as was France, with 1.4 million soldiers dead,[206] not counting other casualties. Germany and Russia were similarly affected.[207]

The French military cemetery with Douaumont ossuary, which contains the remains of more than 130,000 unknown soldiers.

The war had profound economic consequences. Of the 60 million European soldiers who were mobilised from 1914 to 1918, 8 million were killed, 7 million were permanently disabled, and 15 million were seriously injured. Germany lost 15.1% of its active male population, Austria–Hungary lost 17.1%, and France lost 10.5%.[208] About 750,000 German civilians died from starvation caused by the British blockade during the war.[209] By the end of the war, famine had killed approximately 100,000 people in Lebanon.[210] The best estimates of the death toll from the Russian famine of 1921 run from 5 million to 10 million people.[211] By 1922, there were between 4.5 million and 7 million homeless children in Russia as a result of nearly a decade of devastation from World War I, the Russian Civil War, and the subsequent famine of 1920–1922.[212] Numerous anti-Soviet Russians fled the country after the Revolution; by the 1930s the northern Chinese city of Harbin had 100,000 Russians.[213] Thousands more emigrated to France, England, and the United States.

Diseases flourished in the chaotic wartime conditions. In 1914 alone, louse-borne epidemic typhus killed 200,000 in Serbia.[214] From 1918 to 1922, Russia had about 25 million infections and 3 million deaths from epidemic typhus.[215] Whereas before World War I Russia had about 3.5 million cases of malaria, its people suffered more than 13 million cases in 1923.[216] In addition, a major influenza epidemic spread around the world. Overall, the 1918 flu pandemic killed at least 50 million people.[217] [218]

Lobbying by Chaim Weizmann and fear that American Jews would encourage the USA to support Germany culminated in the British government's Balfour Declaration of 1917, endorsing creation of a Jewish homeland in Palestine.[219] A total of more than 1,172,000 Jewish soldiers served in the Allied and Central Power forces in World War I, including 275,000 in Austria-Hungary and 450,000 in Czarist Russia.[220]

The social disruption and widespread violence of the Revolution of 1917 and the ensuing Russian Civil War sparked more than 2,000 pogroms in the former Russian Empire, mostly in the Ukraine.[221] An estimated 60,000–200,000 civilian Jews were killed in the atrocities.[222]

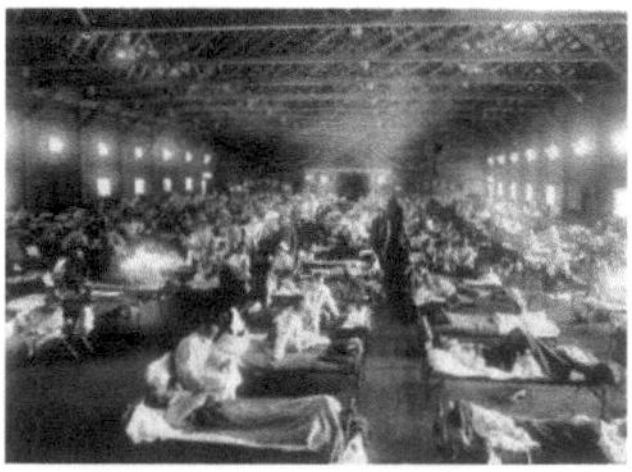

Emergency military hospital during the Spanish flu pandemic, which killed about 675,000 people in the United States alone. Camp Funston, Kansas, 1918

In the aftermath of World War I, Greece fought against Turkish nationalists led by Mustafa Kemal, a war which resulted in a massive population exchange between the two countries under the Treaty of Lausanne.[223] According to various sources,[224] several hundred thousand Pontic Greeks died during this period.[225]

Peace treaties and national boundaries

After the war, the Paris Peace Conference imposed a series of peace treaties on the Central Powers. The 1919 Treaty of Versailles officially ended the war. Building on Wilson's 14th point, the Treaty of Versailles also brought into being the League of Nations on 28 June 1919.[226] [227]

In signing the treaty, Germany acknowledged responsibility for the war, and agreed to pay enormous war reparations and award territory to the victors. The "Guilt Thesis" became a controversial explanation of later events among analysts in Britain and the United States. The Treaty of Versailles caused enormous bitterness in Germany, which nationalist movements, especially the Nazis, exploited with a conspiracy theory they called the *Dolchstosslegende* (Stab-in-the-back legend). The Weimar Republic lost the former colonial possessions and was saddled with accepting blame for the war, as well as paying punitive reparations for it. Unable to pay them with exports (as a result of territorial losses and postwar recession),[228] Germany did so by borrowing from the United States. Runaway inflation in the 1920s contributed to the economic collapse of the Weimar Republic, and the payment of reparations was suspended in 1931 following the Stock Market Crash of 1929 and the beginnings of the Great Depression worldwide.

Greek refugees from Smyrna, Turkey, 1922

Austria–Hungary was partitioned into several successor states, including Austria, Hungary, Czechoslovakia, and Yugoslavia, largely but not entirely along ethnic lines. Transylvania was shifted from Hungary to Greater Romania. The details were contained in the Treaty of Saint-Germain and the Treaty of Trianon. As a result of the Treaty of Trianon, 3.3 million Hungarians came under foreign rule. Although the Hungarians made up 54% of the population of the pre-war Kingdom of Hungary, only 32% of its territory was left to Hungary. Between 1920 and 1924, 354,000 Hungarians fled former Hungarian territories attached to Romania, Czechoslovakia, and Yugoslavia.

The Russian Empire, which had withdrawn from the war in 1917 after the October Revolution, lost much of its western frontier as the newly independent nations of Estonia, Finland, Latvia, Lithuania, and Poland were carved from it. Bessarabia was re-attached to Greater Romania, as it had been a Romanian territory for more than a thousand years.[229]

The Ottoman Empire disintegrated, and much of its non-Anatolian territory was awarded to various Allied powers as protectorates. The Turkish core was reorganised as the Republic of Turkey. The Ottoman Empire was to be partitioned by the Treaty of Sèvres of 1920. This treaty was never ratified by the Sultan and was rejected by the Turkish republican movement, leading to the Turkish Independence War and, ultimately, to the 1923 Treaty of Lausanne.

Legacy

... *"Strange, friend," I said, "Here is no cause to mourn." "None," said the other, "Save the undone years"...*
— Wilfred Owen, *Strange Meeting*, 1918[138]

The first tentative efforts to comprehend the meaning and consequences of modern warfare began during the initial phases of the war, and this process continued throughout and after the end of hostilities.

Memorials

The Beaumont Hamel Newfoundland Memorial in the Somme.

Memorials were erected in thousands of villages and towns. Close to battlefields, those buried in improvised burial grounds were gradually moved to formal graveyards under the care of organisations such as the Commonwealth War Graves Commission, the American Battle Monuments Commission, the German War Graves Commission, and Le Souvenir français. Many of these graveyards also have central monuments to the missing or unidentified dead, such as the Menin Gate memorial and the Thiepval Memorial to the Missing of the Somme.

Surgeon Lt. Col. John McCrae of Canada, author of *In Flanders Fields*, died in 1918 of pneumonia.

On 3 May 1915, during the Second Battle of Ypres, Lieutenant Alexis Helmer was killed. At his graveside, his friend John McCrae, M.D., of Guelph, Ontario, Canada, wrote the memorable poem *In Flanders Fields* as a salute to those who perished in the Great War. Published in *Punch* on 8 December 1915, it is still recited today, especially on Remembrance Day and Memorial Day.[230] [231]

Liberty Memorial in Kansas City, Missouri, is a United States memorial dedicated to all Americans who served in World War I. The site for the Liberty Memorial was dedicated on 1 November 1921. On this day, the supreme Allied commanders spoke to a crowd of more than 100,000 people. It was the only time in history these leaders were together in one place. In attendance were Lieutenant General Baron Jacques of Belgium; General Armando Diaz of Italy; Marshal Ferdinand Foch of France; General Pershing of the United States; and Admiral D. R. Beatty of Great Britain. After three years of construction, the Liberty Memorial was completed and President Calvin Coolidge delivered the dedication speech to a crowd of 150,000 people in 1926.

Liberty Memorial is also home to The National World War I Museum, the only museum dedicated solely to World War I in the United States.

Cultural memory

The First World War had a lasting impact on social memory. It was seen by many in Britain as signalling the end of an era of stability stretching back to the Victorian period, and across Europe many regarded it as a watershed.[232] Historian Samuel Hynes explained:

> A generation of innocent young men, their heads full of high abstractions like Honour, Glory and England, went off to war to make the world safe for democracy. They were slaughtered in stupid battles planned by stupid generals. Those who survived were shocked, disillusioned and embittered by their war experiences, and saw that their real enemies were not the Germans, but the old men at home who had lied to them. They rejected the values of the society that had sent them to war, and in doing so separated their own generation from the past and from their cultural inheritance.[233]

A village war memorial to soldiers killed in World War I.

This has become the most common perception of the First World War, perpetuated by the art, cinema, poems, and stories published subsequently. Films such as *All Quiet on the Western Front*, *Paths of Glory* and *King & Country* have perpetuated the idea, while war-time films including *Camrades*, *Flanders Poppies*, and *Shoulder Arms* indicate that the most contemporary views of the war were overall far more positive.[234] Likewise, the art of Paul Nash, John Nash, Christopher Nevison, and Henry Tonks in Britain painted a negative view of the conflict in keeping with the growing perception, while popular war-time artists such as Muirhead Bone painted more serene and pleasant interpretations subsequently rejected as inaccurate.[233] Several historians have since countered these interpretations:

> These beliefs did not become widely shared because they offered the only accurate interpretation of wartime events. In every respect, the war was much more complicated than they suggest. In recent years, historians have argued persuasively against almost every popular cliché of the First World War. It has been pointed out that, although the losses were devastating, their greatest impact was socially and geographically limited. The many emotions other than horror experienced by soldiers in and out of the front line, including comradeship, boredom, and even enjoyment, have been recognised. The war is not now seen as a 'fight about nothing', but as a war of ideals, a struggle between aggressive militarism and more or less liberal democracy. It has been acknowledged that British generals were often capable men facing difficult challenges, and that it was under their command that the British army played a major part in the defeat of the Germans in 1918: a great forgotten victory.[234]

Siegfried Sassoon (May 1915)

Though these historians have discounted as "myths"[233] [235] these perceptions of the war, they are nevertheless prevalent across much of society. They have dynamically changed according to contemporary influences, reflecting in the 1950s perceptions of the war as 'aimless' following the contrasting Second World War, and emphasising conflict within the ranks during times of class conflict in the 1960s.[234] The majority of additions to the contrary are often rejected.[234]

Social trauma

The social trauma caused by unprecedented rates of casualties manifested itself in different ways, which have been the subject of subsequent historical debate.[236] Some people were revolted by nationalism and its results, and began to work towards a more internationalist world, supporting organisations such as the League of Nations. Pacifism became increasingly popular. Others had the opposite reaction, feeling that only strength and military might could be relied upon in a chaotic and inhumane world. Anti-modernist views were an outgrowth of the many changes taking place in society.

The experiences of the war led to a collective trauma shared by many from all participating countries. The optimism of *la belle époque* was destroyed, and those who had fought in the war were referred to as the Lost Generation.[237] For years afterwards, people mourned the dead, the missing, and the many disabled.[238] Many soldiers returned with severe trauma, suffering from shell shock (also called neurasthenia, a condition related to posttraumatic stress disorder).[239] Many more returned home with few after-effects; however, their silence about the war contributed to the conflict's growing mythological status.[236] In the United Kingdom, mass mobilisation, large casualty rates, and the collapse of the Edwardian era made a strong impression on society. Though many participants did not share in the experiences of combat or spend any significant time at the front, or had positive memories of their service, the images of suffering and trauma became the widely shared perception.[236] Such historians as Dan Todman, Paul Fussell, and Samuel Heyns have all published works since the 1990s arguing that these common perceptions of the war are factually incorrect.[236]

Book distributed by the U.S. War Department to veterans in 1919

Discontent in Germany

The rise of Nazism and fascism included a revival of the nationalist spirit and a rejection of many post-war changes. Similarly, the popularity of the Stab-in-the-back legend (German: *Dolchstoßlegende*) was a testament to the psychological state of defeated Germany and was a rejection of responsibility for the conflict. This conspiracy theory of betrayal became common, and the German populace came to see themselves as victims. The *Dolchstoßlegende*'s popular acceptance in Germany played a significant role in the rise of Nazism. A sense of disillusionment and cynicism became pronounced, with nihilism growing. Many believed the war heralded the end of the world as they had known it because of the high fatalities among a generation of men, the dissolution of governments and empires, and the collapse of capitalism and imperialism.

Communist and socialist movements around the world drew strength from this theory and enjoyed a new level of popularity. These feelings were most pronounced in areas directly or harshly affected by the war. Out of German discontent with the still controversial Treaty of Versailles, Adolf Hitler was able to gain popularity and power.[240] [241] World War II was in part a continuation of the power struggle never fully resolved by the First World War; in fact, it was common for Germans in the 1930s and 1940s to justify acts of international aggression because of perceived injustices imposed by the victors of the First World War.[242] [243] [244]

The establishment of the modern state of Israel and the roots of the continuing Israeli-Palestinian Conflict are partially found in the unstable power dynamics of the Middle East which resulted from World War I.[245] Prior to the end of the war, the Ottoman Empire had maintained a modest level of peace and stability throughout the Middle East.[246] With the fall of the Ottoman government, power vacuums developed and conflicting claims to land and nationhood began to emerge.[247] The political boundaries drawn by the victors of the First World War were quickly imposed, sometimes after only cursory consultation with the local population. In many cases, these continue to be problematic in the 21st-century struggles for national identity.[248] [249] While the dissolution of the Ottoman Empire at the end of World War I was pivotal in contributing to the modern political situation of the Middle East, including the Arab-Israeli conflict,[250] [251] [252] the end of Ottoman rule also spawned lesser known disputes over water and other natural resources.[253]

Further information: Sykes–Picot Agreement

Views in the United States

The announcing of the armistice on November 11, 1918. Philadelphia.

U.S. intervention in the war, as well as the Wilson administration itself, became deeply unpopular. This was reflected in the U.S. Senate's rejection of the Versailles Treaty and membership in the League of Nations. In the interwar era, a consensus arose that U.S. intervention had been a mistake, and the Congress passed laws in an attempt to preserve U.S. neutrality in any future conflict. Polls taken in 1937 and the opening months of World War II established that nearly 60% regarded intervention in WWI as a mistake, with only 28% opposing that view. But, in the period between the fall of France and the attack on Pearl Harbor, public opinion changed dramatically and, for the first time, a narrow plurality rejected the idea that the war had been a mistake.[254]

New national identities

Poland reemerged as an independent country, after more than a century. As a "minor Entente nation" and the country with the most casualties per capita,[255] [256] [257] the Kingdom of Serbia and its dynasty became the backbone of the new multinational state, the Kingdom of Serbs, Croats and Slovenes (later renamed Yugoslavia). Czechoslovakia, combining the Kingdom of Bohemia with parts of the Kingdom of Hungary, became a new nation. Russia became the Soviet Union and lost Finland, Estonia, Lithuania, and Latvia, which became independent countries. The Ottoman Empire was soon replaced by Turkey and several other countries in the Middle East.

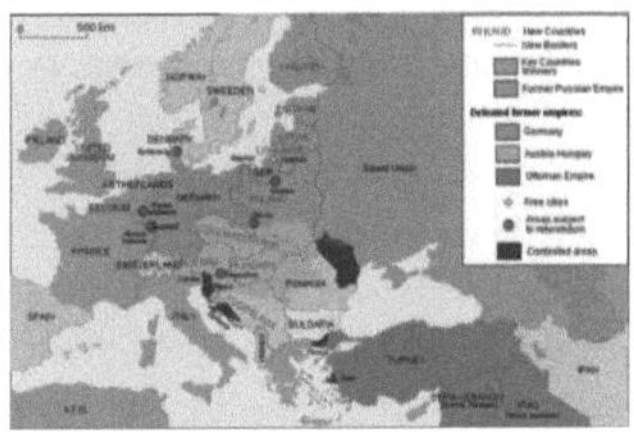

Map of territorial changes in Europe after World War I

In the British Empire, the war unleashed new forms of nationalism. In Australia and New Zealand the Battle of Gallipoli became known as those nations' "Baptism of Fire". It was the first major war in which the newly established countries fought, and it was one of the first times that Australian troops fought as Australians, not just subjects of the British Crown. Anzac Day, commemorating the Australian and New Zealand Army Corps, celebrates this defining moment.[258] [259]

After the Battle of Vimy Ridge, where the Canadian divisions fought together for the first time as a single corps, Canadians began to refer to theirs as a nation "forged from fire".[260] Having succeeded on the same battleground where the "mother countries" had previously faltered, they were for the first time respected internationally for their own accomplishments. Canada entered the war as a Dominion of the British Empire and remained so, although she emerged with a greater measure of independence.[261] [262] When Britain declared war in 1914 the dominions were automatically at war; at the conclusion, Canada, Australia, New Zealand, and South Africa were individual signatories of the Treaty of Versailles.[263]

Economic effects

One of the most dramatic effects of the war was the expansion of governmental powers and responsibilities in Britain, France, the United States, and the Dominions of the British Empire. In order to harness all the power of their societies, governments created new ministries and powers. New taxes were levied and laws enacted, all designed to bolster the war effort; many have lasted to this day. Similarly, the war strained the abilities of some formerly large and bureaucratised governments, such as in Austria–Hungary and Germany; however, any analysis of the long-term effects were clouded by the defeat of these governments.

Germany, 1923: banknotes had lost so much value that they were used as wallpaper. Millions of middle-class Germans were ruined by hyperinflation. When the war began in 1914, a dollar was worth 4.2 marks; by November 1923, the dollar was at 4.2 trillion[264] marks.[265]

Gross domestic product (GDP) increased for three Allies (Britain, Italy, and U.S.), but decreased in France and Russia, in neutral Netherlands, and in the three main Central Powers. The shrinkage in GDP in Austria, Russia, France, and the Ottoman Empire reached 30 to 40%. In Austria, for example, most pigs were slaughtered, so at war's end there was no meat.

In all nations the government's share of GDP increased, surpassing fifty percent in both Germany and France and nearly reaching that level in Britain. To pay for purchases in the United States, Britain cashed in its extensive investments in American railroads and then began borrowing heavily on Wall Street. President Wilson was on the verge of cutting off the loans in late 1916, but allowed a great increase in U.S. government lending to the Allies. After 1919, the U.S. demanded repayment of these loans. The repayments were, in part, funded by German reparations, which, in turn, were supported by American loans to Germany. This circular system collapsed in 1931 and the loans were never repaid. In 1934, Britain owed the US $4.4 billion[266] of World War I debt.[267]

"The Girl Behind the Gun" – women workers, 1915

Macro- and micro-economic consequences devolved from the war. Families were altered by the departure of many men. With the death or absence of the primary wage earner, women were forced into the workforce in unprecedented numbers. At the same time, industry needed to replace the lost labourers sent to war. This aided the struggle for voting rights for women.[268]

In Britain, rationing was finally imposed in early 1918, limited to meat, sugar, and fats (butter and oleo), but not bread. The new system worked smoothly. From 1914 to 1918 trade union membership doubled, from a little over four million to a little over eight million. Work stoppages and strikes became frequent in 1917–1918 as the unions expressed grievances regarding prices, alcohol control, pay disputes, fatigue from overtime and working on Sundays, and inadequate housing.

Britain turned to her colonies for help in obtaining essential war materials whose supply had become difficult from traditional sources. Geologists such as Albert Ernest Kitson were called upon to find new resources of precious minerals in the African colonies. Kitson discovered important new deposits of manganese, used in munitions production, in the Gold Coast.[269]

Article 231 of the Treaty of Versailles (the so-called "war guilt" clause) declared Germany and its allies responsible for all "loss and damage" suffered by the Allies during the war and provided the basis for reparations. The total reparations demanded was 132 billion gold marks, which was far more than the total German gold or foreign exchange. The economic problems that the payments brought, and German resentment at their imposition, are usually cited as one of the more significant factors that led to the end of the Weimar Republic and the beginning of the dictatorship of Adolf Hitler. After Germany's defeat in World War II, payment of the reparations was not resumed. There was, however, outstanding German debt that the Weimar Republic had used to pay the reparations. Germany finished paying off the reparations in October 2010.[270]

See also

- European Civil War
- List of last surviving World War I veterans by country
- List of people associated with World War I
- List of surviving veterans of World War I
- List of wars and anthropogenic disasters by death toll
- Lists of wars
- Lists of World War I topics
- World War I casualties
- World War One – Medal Abbreviations

Notes

[1] Willmott 2003, pp. 10–11
[2] Willmott 2003, p. 15
[3] Keegan 1988, p. 8
[4] Bade & Brown 2003, pp. 167–168
[5] Willmott 2003, p. 307
[6] Taylor 1998, pp. 80–93
[7] Djokić 2003, p. 24
[8] Evans 2004, p. 12
[9] Martel 2003, p. xii ff
[10] Keegan 1988, p. 7
[11] Keegan 1988, p. 11
[12] See "great, adj., adv., and n." in *Oxford English Dictionary* (Second edition, 1989; online version March 2012) (http://www.oed.com/view/Entry/81104)
[13] Shapiro 2006, p. 329 citing a wire service report in *The Indianapolis Star, 20 September 1914*
[14] Keegan 1998, p. 52
[15] Willmott 2003, p. 21
[16] Prior 1999, p. 18
[17] Fromkin 2004, p. 94
[18] Keegan 1998, pp. 48–49
[19] Willmott 2003, pp. 2–23
[20] Willmott 2003, p. 26
[21] Willmott 2003, p. 27
[22] Strachan 2003, p. 68
[23] Willmott 2003, p. 29
[24] "Daily Mirror Headlines: The Declaration of War, Published 4 August 1914" (http://www.bbc.co.uk/history/worldwars/wwone/mirror01_01.shtml). BBC. . Retrieved 9 February 2010.
[25] Strachan 2003, pp. 292–296, 343–354
[26] Farwell 1989, p. 353
[27] Tucker & Roberts 2005, p. 172

[28] John R. Schindler, "Disaster on the Drina: The Austro-Hungarian Army in Serbia, 1914," *War In History* (April 2002) 9#2 pp 159-195 (http://wih.sagepub.com/content/9/2/159.abstract)
[29] Tucker & Roberts 2005, pp. 376–8
[30] Keegan 1968, pp. 224–232
[31] Falls 1960, pp. 79–80
[32] Raudzens 1990, pp. 424
[33] Raudzens 1990, pp. 421–423
[34] Goodspeed 1985, p. 199 (footnote)
[35] Love 1996
[36] Duffy
[37] Tucker & Roberts 2005, p. 1221
[38] Tucker & Roberts 2005, p. 854
[39] Heer 2009, pp. 223–4
[40] Goodspeed 1985, p. 226
[41] Ludendorff 1919, p. 480
[42] Terraine 1963
[43] Perry 1988, p. 27
[44] "Vimy Ridge, Canadian National Memorial" (http://www.ww1westernfront.gov.au/vimy-ridge/index.html), *Australians on the Western Front 1914–1918* (New South Wales Department of Veteran's Affairs and Board of Studies), 2007,
[45] Winegard
[46] Taylor 2007, pp. 39–47
[47] Keene 2006, p. 5
[48] Halpern 1995, p. 293
[49] Zieger 2001, p. 50
[50] Tucker & Roberts 2005, pp. 619–24
[51] Sheffield, Garry, "The First Battle of the Atlantic" (http://www.bbc.co.uk/history/worldwars/wwone/battle_atlantic_ww1_01.shtml), *World Wars In Depth* (BBC), , retrieved 11 November 2009
[52] Gilbert 2004, p. 306
[53] von der Porten 1969
[54] Jones 2001, p. 80
[55] "Nova Scotia House of Assembly Committee on Veterans' Affairs" (http://www.gov.ns.ca/legislature/hansard//comm/va/va_2006nov09.htm), *Hansard*, , retrieved 30 October 2007
[56] Roger Chickering, Stig Förster, Bernd Greiner, German Historical Institute (Washington, D.C.) (2005). " *A world at total war: global conflict and the politics of destruction, 1937–1945* (http://books.google.com/books?id=evVPoSwqrG4C&pg=PA73&dq&hl=en#v=onepage&q=&f=false)". Cambridge University Press. p.73. ISBN 0521834325
[57] Price
[58] Neiberg 2005, pp. 54–55
[59] Tucker & Roberts 2005, pp. 1075–6
[60] Neiberg 2005, pp. 108–10
[61] Tucker, Wood & Murphy 1999, p. 120
[62] *Pyrrhic victory: French strategy and operations in the Great War* (http://books.google.com/?id=vZRmHkdGk44C&pg=PA247&dq=vardar+offensive#v=onepage&q=vardar offensive&f=false), Harvard University Press, 2005;, 2005, p. 491, ISBN 9780674018808, , retrieved 3 October 2010
[63] "The Balkan Front of the World War (in Russian)" (http://militera.lib.ru/h/korsun_ng4/06.html). militera.lib.ru. . Retrieved 27 September 2010.
[64] The Treaty of Alliance Between Germany and Turkey (http://avalon.law.yale.edu/20th_century/turkgerm.asp) 2 August 1914, Yale University
[65] Fromkin 2001, p. 119
[66] Hinterhoff 1984, pp. 499–503
[67] Sachar, pp. 122–138
[68] Gilbert 1994
[69] Page
[70] Giuseppe Praga, Franco Luxardo. *History of Dalmatia*. Giardini, 1993. Pp. 281.
[71] Paul O'Brien. *Mussolini in the First World War: the Journalist, the Soldier, the Fascist*. Oxford, England, UK; New York, New York, USA: Berg, 2005. Pp. 17.
[72] Hickey 2003, pp. 60–65
[73] Tucker 2005, pp. 585–9
[74] "The Battle of Marasti (July 1917)" (http://www.worldwar2.ro/primulrazboi/?language=en&article=116). WorldWar2.ro. 22 July 1917. . Retrieved 8 May 2011.

[75] Cyril Falls, *The Great War*, p. 285

[76] Béla, Köpeczi, *Erdély története* (http://mek.oszk.hu/02100/02109/html/571.html), Akadémiai Kiadó,

[77] Béla, Köpeczi, *History of Transylvania* (http://mek.niif.hu/03400/03407/html/429.html), Akadémiai Kiadó, ISBN 848371020X,

[78] Erlikman, Vadim (2004), *Poteri narodonaseleniia v XX veke : spravochnik*, Moscow, ISBN 5-93165-107-1

[79] Brown 1994, pp. 197–198

[80] Brown 1994, pp. 201–203

[81] *Participants from the Indian subcontinent in the First World War* (http://www.mgtrust.org/ind1.htm), Memorial Gates Trust, , retrieved 12 December 2008

[82] Tucker 2005, p. 715

[83] Meyer 2006, pp. 152–4, 161, 163, 175, 182

[84] Smele

[85] Schindler 2003

[86] Wheeler-Bennett 1956

[87] Mawdsley 2008, pp. 54–55

[88] Kernek 1970, pp. 721–766

[89] Stracham (1998), p. 61

[90] Lyons 1999, p. 243

[91] Marshall, 292.

[92] Heyman 1997, pp. 146–147

[93] Brands 1997, p. 756

[94] Tuchman 1966

[95] Karp 1979

[96] "Woodrow Wilson Urges Congress to Declare War on Germany" (Wikisource)

[97] "Selective Service System: History and Records" (http://www.sss.gov/induct.htm). Sss.gov. . Retrieved 27 July 2010.

[98] Wilgus, p. 52

[99] *Teaching With Documents: Photographs of the 369th Infantry and African Americans during World War I* (http://www.archives.gov/education/lessons/369th-infantry/), U.S. National Archives and Records Administration, , retrieved 29 October 2009

[100] Millett & Murray 1988, p. 143

[101] Kurlander 2006

[102] Shanafelt 1985, pp. 125–30

[103] Westwell 2004

[104] Posen 1984, pp. 190&191

[105] Gray 1991, p. 86

[106] Moon 1996, pp. 495–196

[107] Rickard 2007

[108] Hovannisian, Richard G. (1967). *Armenia on the Road to Independence, 1918*. Berkeley: University of California Press. ISBN 0-5200-0574-0.

[109] See Hovannisian, Richard G. (1971). *The Republic of Armenia: The First Year, 1918–1919*. Berkeley: University of California Press. pp. 1–39. ISBN 0-5200-1805-2.

[110] *The Battle of Amiens: 8 August 1918* (http://www.awm.gov.au/1918/battles/amiens.htm), Australian War Memorial, , retrieved 12 December 2008

[111] *Amiens Map* (http://web.archive.org/web/20070617055415/http://www.awm.gov.au/1918/battles/amiensmap.htm), Australian War Memorial, archived from the original (http://www.awm.gov.au/1918/battles/amiensmap.htm) on 17 June 2007, , retrieved 24 October 2009 (archived 17 June 2007)

[112] Rickard 2001

[113] Pitt 2003

[114] Maurice 1918

[115] Gray & Argyle 1990

[116] Nicholson 1962

[117] Ludendorff 1919

[118] Jenkins 2009, p. 215

[119] McLellan, p. 49

[120] Gibbs 1918b

[121] Gibbs 1918a

[122] Stevenson 2004, p. 380

[123] Hull 2006, pp. 307–10

[124] Stevenson 2004, p. 383

[125] Stevenson 2004

[126] (in French) *Clairière de l'Armistice* (http://www.compiegne.fr/decouvrir/clairierearmistice.asp), Ville de Compiègne, , retrieved 3 December 2008

[127] "1918 Timeline" (http://www.indiana.edu/~league/1918.htm). *League of Nations Photo Archive*. . Retrieved 20 November 2009.

[128] Lindsay, Robert, "The Last Hours" (http://www.nwbattalion.com/last.html), *28th (Northwest) Battalion Headquarters*, , retrieved 20 November 2009

[129] Gunther, Henry (Wednesday 29, 2008), *[[BBC News|BBC Magazine* (http://news.bbc.co.uk/2/hi/uk_news/magazine/7696021.stm)*]], , retrieved 6 December 2012*

[130] Tomas (15 February 2010), *[[11 Facts about the End of the Great War* (http://listverse.com/2010/02/15/11-facts-about-the-end-of-the-great-war/)*]], , retrieved 6 December 2012*

[131] Baker 2006

[132] Chickering 2004, pp. 185–188

[133] Hartcup 1988, p. 154

[134] Hartcup 1988, pp. 82–86

[135] Mosier 2001, pp. 42–48

[136] Harcup 1988

[137] Raudzens, p. 421

[138] *Wilfred Owen: poems*, (Faber and Faber, 2004)

[139] Raudzens

[140] Heller 1984

[141] Postwar pulp novels on future "gas wars" included Reginald Glossop's 1932 novel *Ghastly Dew* and Neil Bell's 1931 novel *The Gas War of 1940.*

[142] Eric Lawson, Jane Lawson (2002). " *The First Air Campaign: August 1914– November 1918* (http://books.google.com/books?id=9PGHckhHiX0C&pg=PT123&dq&hl=en#v=onepage&q=&f=false)". Da Capo Press. p.123. ISBN 0306812134

[143] Cross 1991

[144] Cross 1991, pp. 56–57

[145] Winter 1983

[146] Johnson 2001

[147] Price 1980

[148] International Association of Genocide Scholars (13 June 2005). "Open Letter to the Prime Minister of Turkey Recep Tayyip Erdoğan" (http://web.archive.org/web/20071006024502/http://www.genocidewatch.org/TurkishPMIAGSOpenLetterreArmenia6-13-05.htm). Genocide Watch (via archive.org). Archived from the original (http://www.genocidewatch.org/TurkishPMIAGSOpenLetterreArmenia6-13-05.htm) on 6 October 2007. .

[149] Lewy 2005, p. 57

[150] Ferguson 2006, p. 177

[151] Balakian 2003, pp. 195–196

[152] Fromkin 1989, pp. 212–215

[153] (PDF) *Resolution on genocides committed by the Ottoman empire* (http://www.genocidescholars.org/images/Resolution_on_genocides_committed_by_the_Ottoman_Empire.pdf), International Association of Genocide Scholars,

[154] Gaunt, David. *Massacres, Resistance, Protectors: Muslim-Christian Relations in Eastern Anatolia during World War I* (http://books.google.se/books?id=4mug9LrpLKcC&printsec=frontcover&dq=Massacres,+Resistance,+Protectors&cd=1#v=onepage&q=&f=false). Piscataway, New Jersey: Gorgias Press, 2006.

[155] Schaller, Dominik J; Zimmerer, Jürgen (2008). "Late Ottoman genocides: the dissolution of the Ottoman Empire and Young Turkish population and extermination policies – introduction". *Journal of Genocide Research* **10** (1): 7–14. doi:10.1080/14623520801950820.

[156] *A People on the Move: Germans in Russia and in the Former Soviet Union: 1763–1997* (http://lib.ndsu.nodak.edu/grhc/history_culture/history/people.html), North Dakota State University Libraries, , retrieved 17 November 2009

[157] *WWI and the Jews* (http://www.myjewishlearning.com/history/Modern_History/1914-1948/WWI_and_the_Jews.shtml), MyJewishLearning.com, , retrieved 17 November 2009

[158] *Timeline 1900s* (http://www.loc.gov/exhibits/haventohome/timeline/haven-timeline_3.html), The Library of Congress,

[159] *The Germans from Russia: Children of the Steppe/Children of the Prairie* (http://archive.prairiepublic.org/features/GFR/timeline.htm), Prairie Public Broadcasting, , retrieved 17 November 2009

[160] "Pogroms" (http://www.jewishvirtuallibrary.org/jsource/judaica/ejud_0002_0016_0_15895.html), *Encyclopaedia Judaica* (Jewish Virtual Library), , retrieved 17 November 2009

[161] *Jewish Modern and Contemporary Periods (ca. 1700–1917)* (http://www.jewishvirtuallibrary.org/jsource/History/modtimeline.html), Jewish Virtual Library, , retrieved 17 November 2009

[162] John Horne and Alan Kramer, *German Atrocities, 1914: A History of Denial* (Yale U.P. 2001) ch 1-2, esp. p. 76

[163] Horne and Kramer, *German Atrocities, 1914: A History of Denial* ch 3-4 show there were no "franc-tireurs" in Belgium.

[164] Horne and Kramer, *German Atrocities, 1914: A History of Denial* ch 5-8

[165] Keegan 1998, pp. 82–83

[166] *Forgotten Voices of the Great War* (http://www.forgottenvoices.co.uk/), Imperial War Museum, , retrieved 30 March 2008

[167] Phillimore & Bellot 1919, pp. 4–64
[168] Ferguson 1999, pp. 368–9
[169] Blair 2005
[170] Cook 2006, pp. 637&-665
[171] Speed 1990
[172] Ferguson 1999
[173] Morton 1992
[174] Bass 2002, p. 107
[175] *The Mesopotamia campaign* (http://www.nationalarchives.gov.uk/pathways/firstworldwar/battles/mesopotamia.htm), British National Archives, , retrieved 10 March 2007
[176] "Prisoners of Turkey: Men of Kut *Driven along like beasts*" (http://www.awm.gov.au/stolenyears/ww1/turkey/story2.asp), *Stolen Years: Australian Prisoners of War* (Australian War Memorial), , retrieved 10 December 2008
[177] "ICRC in WWI: overview of activities" (http://www.icrc.org/Web/Eng/siteeng0.nsf/html/57JQGQ). Icrc.org. . Retrieved 15 June 2010.
[178] Monday, 1 Sep. 1924 (1 September 1924). "GERMANY: Notes, Sep. 1, 1924" (http://www.time.com/time/magazine/article/0,9171,768983,00.html). *Time*. . Retrieved 15 June 2010.
[179] Fortescue 28 October 1915, p. 1
[180] *Granville Roland Fortescue* (http://www.arlingtoncemetery.net/fortesc.htm), Arlington National Cemetery, , retrieved 17 November 2009
[181] Sisemore 2003
[182] Tucker & Roberts 2005, p. 1189
[183] Tucker & Roberts 2005, p. 117
[184] Tucker & Roberts 2005, p. 335
[185] Tucker & Roberts 2005, p. 219
[186] Tucker & Roberts 2005, p. 1001
[187] Tucker & Roberts 2005, p. 1069
[188] Tucker & Roberts 2005, p. 884
[189] Tucker & Roberts 2005, p. 209
[190] Tucker & Roberts 2005, p. 596
[191] Tucker & Roberts 2005, p. 826
[192] Dennis Mack Smith. 1997. *Modern Italy; A Political History*. Ann Arbor: The University of Michigan Press. Pp. 284.
[193] Roger Aubert, et al.; Margit Resch (tr.), John Dolan (1981). "Chapter 37: The Outbreak of World War I". *History of the Church*. **9**. London: Burns & Oates. p. 521. ISBN 0860120910.
[194] "Who's Who — Pope Benedict XV" (http://www.firstworldwar.com/bio/popebenedict.htm). firstworldwar.com. Saturday, 22 August 2009. .
[195] *"Merely For the Record": The Memoirs of Donald Christopher Smith 1894–1980*. By Donald Christopher Smith. Edited by John William Cox, Jr. Bermuda.
[196] Tucker & Roberts 2005, p. 584
[197] Tucker & Roberts 2005, p. 586
[198] Lehmann 1999, p. 62
[199] Uzbeks (http://www.country-data.com/cgi-bin/query/r-12499.html). Based on the Country Studies Series by Federal Research Division of the Library of Congress.
[200] Seton-Watson, Christopher. 1967. *Italy from Liberalism to Fascism: 1870 to 1925*. London: Methuen & Co. Ltd. Pp. 471
[201] " The Conscription Crisis (http://history.cbc.ca/history/?MIval=EpisContent.html&series_id=1&episode_id=12&chapter_id=2&page_id=3&lang=E)". CBC.ca.
[202] *World War I* (http://www.britannica.com/EBchecked/topic/91513/Canada/43004/World-War-I), Encyclopædia Britannica, , retrieved 5 December 2009.
[203] " Charles (I) (emperor of Austria) (http://www.britannica.com/EBchecked/topic/106679/Charles-I#ref214141)". "Encyclopædia Britannica."
[204] Cockfield 1997, pp. 171–237
[205] Havighurst 1985, p. 131
[206] "France's oldest WWI veteran dies" (http://news.bbc.co.uk/2/hi/europe/7199127.stm), *BBC News*, 20 Jan 2008.
[207] Spencer Tucker (2005), *Encyclopedia of World War I* (http://books.google.com/books?id=2YqjfHLyyj8C&pg=PA273&dq&hl=en#v=onepage&q=&f=false), ABC-CLIO, p. 273. ISBN 1851094202
[208] Kitchen 2000, p. 22
[209] "Lebensmittelversorgung" (http://www.dhm.de/lemo/html/wk1/wirtschaft/versorgung/index.html) (in German), *LeMO: Lebendiges virtuelles Museum Online* (German Historical Museum), ISBN 3515048057, , retrieved 12 November 2009, "Die miserable Versorgung mit Lebensmitteln erreichte 1916/17 im "Kohlrübenwinter" einen dramatischen Höhepunkt. Während des Ersten Weltkriegs starben in Deutschland rund 750.000 Menschen an Unterernährung und an deren Folgen."

[210] Saadi
[211] "Food as a Weapon" (http://www.hoover.org/publications/digest/6731711.html), *Hoover Digest* (Hoover Institution),
[212] Ball 1996, pp. 16, 211
[213] *The Russians are coming (Russian influence in Harbin, Manchuria, China; economic relations)* (http://findarticles.com/p/articles/mi_hb5037/is_199501/ai_n18298515/), The Economist (US), January 1995, , retrieved 17 November 2009
[214] Tschanz
[215] Conlon
[216] William Hay Taliaferro, *Medicine and the War* (http://books.google.com/books?id=HcOAnAINJZAC&pg=PA65&dq&hl=en#v=onepage&q=&f=false),(1972), p.65. ISBN 0836926293
[217] Knobler 2005
[218] *Influenza Report* (http://www.influenzareport.com/ir/overview.htm), , retrieved 17 November 2009
[219] "Balfour Declaration" (United Kingdom 1917) (http://www.britannica.com/EBchecked/topic/50162/Balfour-Declaration), *Encyclopædia Britannica.*
[220] " The Jewish Agency for Israel Timeline (http://www.jewishagency.org/JewishAgency/English/Jewish+Education/Compelling+Content/Jewish+History/Zionist+Institutions/JAFI+Timeline/1917-1919.htm)"
[221] "Pogroms" (http://www.jewishvirtuallibrary.org/jsource/judaica/ejud_0002_0016_0_15895.html), *Encyclopaedia Judaica*, , retrieved 17 November 2009
[222] "Jewish Modern and Contemporary Periods (ca. 1700–1917)" (http://www.jewishvirtuallibrary.org/jsource/History/modtimeline.html), *Jewish Virtual Library*, , retrieved 17 November 2009
[223] "The Diaspora Welcomes the Pope" (http://www.spiegel.de/international/0,1518,451140,00.html), *Der Spiegel* Online. 28 November 2006.
[224] R. J. Rummel, "The Holocaust in Comparative and Historical Perspective," 1998, *Idea Journal of Social Issues*, Vol.3 no.2
[225] Chris Hedges, "A Few Words in Greek Tell of a Homeland Lost" (http://www.nytimes.com/2000/09/17/nyregion/a-few-words-in-greek-tell-of-a-homeland-lost.html), *The New York Times*, 17 Sep 2000
[226] Magliveras 1999, pp. 8–12
[227] Northedge 1986, pp. 35–36
[228] Keynes 1920
[229] Clark 1927
[230] *John McCrae* (http://www.histori.ca/minutes/minute.do?id=10200), Historica,
[231] Evans David, "John McCrae" (http://www.thecanadianencyclopedia.com/index.cfm?PgNm=TCE&ArticleId=A0004849), *Canadian Encyclopedia*,
[232] Mark David Sheftall, *Altered Memories of the Great War: Divergent Narratives of Britain, Australia, New Zealand, and Canada* (2010)
[233] Hynes, Samuel Lynn (1991), *A war imagined: the First World War and English culture*, Atheneum, pp. i–xii, ISBN 9780689121289
[234] Todman, Daniel (2005), *The Great War: myth and memory*, Hambledon and London, pp. 153–221, ISBN 9781852854591
[235] Fussell, Paul (2000), *The Great War and modern memory* (http://books.google.com/?id=D9iNQYfeKdwC), Oxford University Press US, pp. 1–78, ISBN 9780195133325, , retrieved 18 May 2010
[236] Todman, D. *The Great War, Myth and Memory*, p. xi–xv.
[237] Roden
[238] Wohl 1979
[239] Tucker & Roberts 2005, pp. 108–1086
[240] *The Ending of World War One, and the Legacy of Peace* (http://www.bbc.co.uk/history/worldwars/wwone/war_end_01.shtml), BBC,
[241] *The Rise of Hitler* (http://www.schoolshistory.org.uk/hitlergainspower.htm), , retrieved 12 November 2009
[242] "World War II" (http://web.archive.org/web/20080704030736/http://www.britannica.com/eb/article-9110199/World-War-II), *Britannica Online Encyclopedia* (Encyclopædia Britannica Inc.), archived from the original (http://www.britannica.com/eb/article-9110199/World-War-II) on 4 July 2008, , retrieved 12 November 2009
[243] Baker, Kevin (June 2006), "Stabbed in the Back! The past and future of a right-wing myth" (http://harpers.org/StabbedInTheBack.html), *Harper's Magazine*,
[244] Chickering 2004
[245] Economist 2005
[246] Hooker 1996
[247] Muller 2008
[248] Kaplan 1993
[249] Salibi 1993
[250] Evans 2005
[251] Israeli Foreign Ministry
[252] Gelvin 2005
[253] Isaac & Hosh 1992
[254] "1941 Gallup poll" (http://news.google.com/newspapers?id=RPUaAAAAIBAJ&sjid=YEwEAAAAIBAJ&pg=3742,5638207&dq=war+poll&hl=en). Google. . Retrieved 15 June 2010.

[255] "Appeals to Americans to Pray for Serbians" (http://query.nytimes.com/mem/archive-free/pdf?_r=1&res=9406E4D8143EE433A25754C2A9619C946996D6CF). *The New York Times*. 27 July 1918. .

[256] "Serbia Restored" (http://query.nytimes.com/mem/archive-free/pdf?_r=1&res=990CEFDC113BEE3ABC4D53DFB7678383609EDE). *The New York Times*. 5 November 1918. .

[257] Simpson, Matt (22 August 2009). "The Minor Powers During World War One – Serbia" (http://www.firstworldwar.com/features/minorpowers_serbia.htm). firstworldwar.com. .

[258] "'ANZAC Day' in London; King, Queen, and General Birdwood at Services in Abbey" (http://query.nytimes.com/gst/abstract.html?res=9400E1DD113FE233A25755C2A9629C946796D6CF&scp=12&sq=New+Zealand+anzac&st=p). *The New York Times*. 26 April 1916. .

[259] *The ANZAC Day tradition* (http://www.awm.gov.au/commemoration/anzac/anzac_tradition.asp), Australian War Memorial, , retrieved 2 May 2008

[260] *Vimy Ridge* (http://www.warmuseum.ca/cwm/exhibitions/guerre/vimy-ridge-e.aspx), Canadian War Museum, , retrieved 22 October 2008

[261] *The War's Impact on Canada* (http://www.warmuseum.ca/cwm/exhibitions/guerre/war-impact-e.aspx), Canadian War Museum, , retrieved 22 October 2008

[262] *Canada's last WW1 vet gets his citizenship back* (http://www.cbc.ca/canada/ottawa/story/2008/05/09/babcock-citizen.html), CBC News, 9 May 2008,

[263] Documenting Democracy (http://foundingdocs.gov.au/item-did-23.html)], retrieved 2012-03-31

[264] 10^{12} in this context – see Long and short scales

[265] " Germany in the Era of Hyperinflation (http://www.spiegel.de/international/germany/0,1518,641758,00.html)". Spiegel Online. 14 August 2009.

[266] 10^{9} in this context – see Long and short scales

[267] " What's a little debt between friends? (http://news.bbc.co.uk/2/hi/uk_news/magazine/4757181.stm)". BBC News. 10 May 2006.

[268] Noakes, Lucy (2006). *Women in the British Army: war and the gentle sex, 1907–1948*. Abingdon, England: Routledge. p. 48. ISBN 0-415-39056-7.

[269] Green 1938, pp. CXXVI

[270] "Germany finishes paying WWI reparations, ending century of 'guilt'" (http://www.csmonitor.com/World/Europe/2010/1004/Germany-finishes-paying-WWI-reparations-ending-century-of-guilt). . Christian Science Monitor. 4 October 2010.

References

For a comprehensive bibliography see List of books about World War I

- *American Armies and Battlefields in Europe: A History, Guide, and Reference Book* (http://www.secstate.wa.gov/history/ww1/maps.aspx), U.S. Government Printing Office, 1938, OCLC 59803706
- *Army Art of World War I* (http://www.secstate.wa.gov/history/publications_detail.aspx?p=28), United States Army Center of Military History: Smithsonian Institution, National Museum of American History, 1993, OCLC 28608539
- Asghar, Syed Birjees (12 June 2005), *A Famous Uprising* (http://web.archive.org/web/20070803144344/http://www.dawn.com/weekly/dmag/archive/050612/dmag14.htm), Dawn Group, archived from the original (http://www.dawn.com/weekly/dmag/archive/050612/dmag14.htm) on 3 August 2007, retrieved 2 November 2007
- Ashworth, Tony (2000) [1980], *Trench warfare, 1914–18 : the live and let live system*, London: Pan, ISBN 0330480685, OCLC 247360122
- Bade, Klaus J; Brown, Allison (tr.) (2003), *Migration in European History*, The making of Europe, Oxford: Blackwell, ISBN 0631189394, OCLC 52695573 (translated from the German)
- Baker, Kevin (June 2006), "Stabbed in the Back! The past and future of a right-wing myth", *Harper's Magazine*
- Balakian, Peter (2003), *The Burning Tigris: The Armenian Genocide and America's Response*, New York: HarperCollins, ISBN 9780060198404, OCLC 56822108
- Ball, Alan M (1996), *And Now My Soul Is Hardened: Abandoned Children in Soviet Russia, 1918–1930*, Berkeley: University of California Press, ISBN 9780520206946, reviewed in Hegarty, Thomas J (March–June 1998), "And Now My Soul Is Hardened: Abandoned Children in Soviet Russia, 1918–1930" (http://findarticles.com/p/articles/mi_qa3763/is_/ai_n8801575), *Canadian Slavonic Papers*

- Bass, Gary Jonathan (2002), *Stay the Hand of Vengeance: The Politics of War Crimes Tribunals*, Princeton, New Jersey: Princeton University Press, pp. 424pp, ISBN 0691092788, OCLC 248021790
- Blair, Dale (2005), *No Quarter: Unlawful Killing and Surrender in the Australian War Experience, 1915–1918*, Charnwood, Australia: Ginninderra Press, ISBN 1740272919, OCLC 62514621
- Brands, Henry William (1997), *T. R.: The Last Romantic*, New York: Basic Books, ISBN 0465069584, OCLC 36954615
- Brown, Judith M. (1994), *Modern India: The Origins of an Asian Democracy*, Oxford and New York: Oxford University Press. Pp. xiii, 474, ISBN 0198731132.
- Chickering, Rodger (2004), *Imperial Germany and the Great War, 1914–1918*, Cambridge: Cambridge University Press, ISBN 0521839084, OCLC 55523473
- Clark, Charles Upson (1927), *Bessarabia, Russia and Roumania on the Black Sea* (http://depts.washington.edu/cartah/text_archive/clark/meta_pag.shtml), New York: Dodd, Mead, OCLC 150789848
- Cockfield, Jamie H (1997), *With snow on their boots : The tragic odyssey of the Russian Expeditionary Force in France during World War 1*, Palgrave Macmillan, ISBN 0312220820
- Conlon, Joseph M (PDF), *The historical impact of epidemic typhus* (http://entomology.montana.edu/historybug/TYPHUS-Conlon.pdf), Montana State University, retrieved 21 April 2009
- Cook, Tim (2006), "The politics of surrender: Canadian soldiers and the killing of prisoners in the First World War", *The Journal of Military History* **70** (3): 637–665, doi:10.1353/jmh.2006.0158
- Cross, Wilbur L (1991), *Zeppelins of World War I*, New York: Paragon Press, ISBN 9781557783820, OCLC 22860189
- Djokić, Dejan (2003), *Yugoslavism : histories of a failed idea, 1918–1992*, London: Hurst, OCLC 51093251
- Dignan, Don K (February 1971), "The Hindu Conspiracy in Anglo-American Relations during World War I", *The Pacific Historical Review* (University of California Press) **40** (1): 57–76, ISSN 0030-8684, JSTOR 3637829
- Doughty, Robert A. (2005), *Pyrrhic victory: French strategy and operations in the Great War* (http://books.google.com/books?id=vZRmHkdGk44C), Harvard University Press, ISBN 9780674018808
- Duffy, Michael, *Somme* (http://www.firstworldwar.com/battles/somme.htm), First World War.com, ISBN 0297846892, retrieved 25 February 2007
- Evans, David (2004), *The First World War*, Teach yourself, London: Hodder Arnold, ISBN 0340884894, OCLC 224332259
- Evans, Leslie (27 May 2005), *Future of Iraq, Israel-Palestine Conflict, and Central Asia Weighed at International Conference* (http://www.international.ucla.edu/article.asp?parentid=24920), UCLA International Institute, retrieved 30 December 2008
- Falls, Cyril Bentham (1960), *The First World War*, London: Longmans, ISBN 1843422727, OCLC 460327352
- Farwell, Byron (1989), *The Great War in Africa, 1914–1918*, W.W. Norton, ISBN 9780393305647
- Ferguson, Niall (1999), *The Pity of War*, New York: Basic Books, pp. 563pp, ISBN 046505711X, OCLC 41124439
- Ferguson, Niall (2006), *The War of the World: Twentieth-Century Conflict and the Descent of the West*, New York: Penguin Press, ISBN 1594201005
- Findley, Carter Vaughn; Rothney, J.A. (2006), *Twentieth Century World* (6th ed.), Boston: Houghton Mifflin
- Fortescue, Granville Roland (28 October 1915), "London in Gloom over Gallipoli; Captain Fortescue in Book and Ashmead-Bartlett in Lecture Declare Campaign Lost. Say Allies Can't Advance; Attack on Allied Diplomacy in Correspondent's Doleful Talk Passed by Censor" (http://query.nytimes.com/gst/abstract.html?res=9907E3DE1E38E633A2575BC2A9669D946496D6CF), *New York Times*
- Fraser, Thomas G (April 1977), "Germany and Indian Revolution, 1914–18", *Journal of Contemporary History* (Sage Publications) **12** (2): 255–272, doi:10.1177/002200947701200203, ISSN 0022-0094
- Fromkin, David (2001), *A Peace to End All Peace: The Fall of the Ottoman Empire and the Creation of the Modern Middle East*, New York: Owl Books, p. 119, ISBN 0805068848, OCLC 53814831

- Fromkin, David (2004), *Europe's Last Summer: Who Started the Great War in 1914?*, New York: Alfred A. Knopf, ISBN 0375411569, OCLC 53937943
- Gelvin, James L (2005), *The Israel-Palestine Conflict: One Hundred Years of War*, Cambridge: Cambridge University Press, ISBN 0521852897, OCLC 59879560
- Gibbs, Phillip (26 October 1918 published 30 October 1918), "Fall of Ghent Near, German Flank in Peril" (http://query.nytimes.com/gst/abstract.html?res=9F05E4D61539E13ABC4850DFB6678383609EDE&scp=4&sq=Ghent+1918&st=p), *New York Times*
- Gibbs, Phillip (15 November 1918), "Ghent Burghers Hail Liberators" (http://query.nytimes.com/mem/archive-free/pdf?res=940DE1DC1239E13ABC4D52DFB7678383609EDE) (PDF), *New York Times*
- Gray, Randal; Argyle, Christopher (1990), *Chronicle of the First World War*, New York: Facts on File, ISBN 9780816025954, OCLC 19398100
- Gilbert, Martin (2004), *The First World War: A Complete History*, Clearwater, Florida: Owl Books, p. 306, ISBN 0805076174, OCLC 34792651
- Goodspeed, Donald James (1985), *The German Wars 1914–1945*, New York: Random House; Bonanza, ISBN 9780517467909
- Gray, Randal (1991), *Kaiserschlacht 1918: the final German offensive*, Osprey, ISBN 9781855321571
- Green, John Frederick Norman (1938), "Obituary: Albert Ernest Kitson", *Geological Society Quarterly Journal* (Geological Society) **94**
- Haber, Lutz Fritz (1986), *The Poisonous Cloud: Chemical Warfare in the First World War*, Oxford: Clarendon, ISBN 0198581424, OCLC 12051072
- Halpern, Paul G (1995), *A Naval History of World War I*, New York: Routledge, ISBN 1857284984, OCLC 60281302
- Harrach, Franz, "Archduke Franz Ferdinand's Assassination, 28 June 1914: Memoir of Count Franz von Harrach", *Primary Documents* (First World War.com)
- Hartcup, Guy (1988), *The War of Invention; Scientific Developments, 1914–18*, Brassey's Defence Publishers, ISBN 0-08-033591-8
- Havighurst, Alfred F (1985), *Britain in transition: the twentieth century* (4 ed.), University of Chicago Press, ISBN 9780226319711
- Heer, Germany (2009), *German and Austrian Tactical Studies*, ISBN 9781110765164
- Heller, Charles E (1984), *Chemical warfare in World War I : the American experience, 1917–1918* (http://www-cgsc.army.mil/carl/resources/csi/Heller/HELLER.asp), Fort Leavenworth, Kansas: Combat Studies Institute, OCLC 123244486
- Herbert, Edwin (2003), *Small Wars and Skirmishes 1902–1918: Early Twentieth-century Colonial Campaigns in Africa, Asia and the Americas*, Nottingham: Foundry Books Publications, ISBN 1901543056
- Heyman, Neil M (1997), *World War I*, Guides to historic events of the twentieth century, Westport, Connecticut: Greenwood Press, ISBN 0313298807, OCLC 36292837
- Hickey, Michael (2003), *The Mediterranean Front 1914–1923*, The First World War, **4**, New York: Routledge, pp. 60–65, ISBN 0415968445, OCLC 52375688
- Hinterhoff, Eugene (1984), Young, Peter, ed., "The Campaign in Armenia", *Marshall Cavendish Illustrated Encyclopedia of World War I* (New York: Marshall Cavendish) **ii**, ISBN 0863071813
- Hooker, Richard (1996), *The Ottomans* (http://www.wsu.edu/~dee/OTTOMAN/OTTOMAN1.HTM), Washington State University, retrieved 30 December 2008
- Hoover, Herbert; Wilson, Woodrow (1958), *Ordeal of Woodrow Wilson*, New York: McGraw-Hill, ISBN 0943875412, OCLC 254607345
- Hughes, Thomas L (October 2002), "The German Mission to Afghanistan, 1915–1916", *German Studies Review* (German Studies Association) **25** (3): 447–476, doi:10.2307/1432596, ISSN 0149-7952, JSTOR 1432596

- Hull, Isabel Virginia (2006), *Absolute destruction: military culture and the practices of war in Imperial Germany*, Cornell University Press, ISBN 9780801472930
- Isaac, Jad; Hosh, Leonardo (7–9 May 1992), *Roots of the Water Conflict in the Middle East* (http://web.archive.org/web/20060928053605/http://www.oranim.ac.il/courses/meast/water/Roots+of+the+Water+Conflict+in+the+Middle+East.htm), University of Waterloo, archived from the original (http://www.oranim.ac.il/courses/meast/water/Roots+of+the+Water+Conflict+in+the+Middle+East.htm) on 28 September 2006
- Jenkins, Burris A (2009), *Facing the Hindenburg Line*, BiblioBazaar, ISBN 9781110812387
- Johnson, Douglas Wilson (1921), *Battlefields of the World War, Western and Southern Fronts* (http://openlibrary.org/b/OL23383739M/Battlefields_of_the_World_War_western_and_southern_fronts), New York: Oxford University Press, ISBN 1432637398, OCLC 688071
- Johnson, James Edgar (2001), *Full Circle: The Story of Air Fighting*, London: Cassell, ISBN 0304358606, OCLC 45991828
- Jones, Howard (2001), *Crucible of Power: A History of U.S. Foreign Relations Since 1897*, Wilmington, Delaware: Scholarly Resources Books, ISBN 0842029184, OCLC 46640675
- Kaplan, Robert D (February 1993), "Syria: Identity Crisis" (http://www.theatlantic.com/doc/199302/kaplan), *The Atlantic*, retrieved 30 December 2008
- Karp, Walter (1979), *The Politics of War* (1st ed.), ISBN 006012265X, OCLC 4593327, Wilson's maneuvering U.S. into war
- Keegan, John (1998), *The First World War*, Hutchinson, ISBN 0091801788, general military history
- Keene, Jennifer D (2006), *World War I*, Westport, Connecticut: Greenwood Press, p. 5, ISBN 0313331812, OCLC 70883191
- Kennedy, David M (2004), *Over here: the First World War and American society*, Oxford University Press, ISBN 9780195173994
- Kernek, Sterling (December 1970), "The British Government's Reactions to President Wilson's 'Peace' Note of December 1916", *The Historical Journal* **13** (4): 721–766, JSTOR 2637713
- Keynes, John Maynard (1920), *The Economic Consequences of the Peace*, New York: Harcourt, Brace and Howe, ISBN 0521220955, OCLC 213487540
- Kitchen, Martin (2000) [1980], *Europe Between the Wars*, New York: Longman, ISBN 0582418690, OCLC 247285240
- Knobler, Stacey L, ed. (2005), *The Threat of Pandemic Influenza: Are We Ready? Workshop Summary* (http://www.nap.edu/books/0309095042/html/7.html), Washington DC: National Academies Press, ISBN 0309095042, OCLC 57422232
- Kurlander, Eric (2006) (Book review), *Steffen Bruendel. Volksgemeinschaft oder Volksstaat: Die "Ideen von 1914" und die Neuordnung Deutschlands im Ersten Weltkrieg* (http://www.h-net.org/reviews/showrev.cgi?path=101921145898314), H-net, retrieved 17 November 2009
- Lehmann, Hartmut; van der Veer, Peter, eds. (1999), *Nation and religion: perspectives on Europe and Asia*, Princeton, New Jersey: Princeton University Press, ISBN 0691012326, OCLC 39727826
- Lewy, Guenter (2005), *The Armenian Massacres in Ottoman Turkey: A Disputed Genocide*, Salt Lake City, Utah: University of Utah Press, ISBN 0874808499, OCLC 61262401
- Love, Dave (May 1996), "The Second Battle of Ypres, April 1915" (http://www.worldwar1.com/sf2ypres.htm), *Sabretasche* **26** (4)
- Lyons, Michael J (1999), *World War I: A Short History* (2nd ed.), Prentice Hall, ISBN 0130205516
- Ludendorff, Erich (1919), *My War Memories, 1914–1918*, OCLC 60104290 also published by Harper as "Ludendorff's Own Story, August 1914 – November 1918: The Great War from the Siege of Liege to the Signing of the Armistice as Viewed from the Grand Headquarters of the German Army" OCLC 561160 (original title *Meine Kriegserinnerungen, 1914–1918*)

- Magliveras, Konstantinos D (1999), *Exclusion from Participation in International Organisations: The Law and Practice behind Member States' Expulsion and Suspension of Membership*, Martinus Nijhoff Publishers, ISBN 9041112391
- Maurice, Frederick Barton (18 August 1918), "Foe's reserves now only 16 divisions; Allies' Counteroffensive has reduced them from 60, Gen. Maurice says Ludendorff in dilemma; he must choose between giving up offensive projects and shortening his line" (http://query.nytimes.com/gst/abstract.html?res=9B02EFD6103BEE3ABC4052DFBE668383609EDE&scp=8&sq=Ludendorff+Amiens+1918&st=p), *New York Times*
- Martel, Gordon (2003), *The Origins of the First World War*, Pearson Longman, Harlow
- Mawdsley, Evan (2008), *The Russian Civil War* (Edinburgh ed.), Birlinn location, ISBN 1843410419
- McDermott, T. P., *USA's Boy Scouts and World War I Liberty Loan Bonds* (http://www.sossi.org/journal/scouts-ww1-liberty-bonds.pdf)
- McLellan, Edwin N, *The United States Marine Corps in the World War* (http://www.ibiblio.org/hyperwar/AMH/XX/WWI/USMC/USMC-WWI.html#XIV)
- Meyer, Gerald J (2006), *A World Undone: The Story of the Great War 1914 to 1918*, Random House, ISBN 9780553803549
- Millett, Allan Reed; Murray, Williamson (1988), *Military Effectiveness*, Boston: Allen Unwin, ISBN 0044450532, OCLC 220072268
- Moon, John Ellis van Courtland (July 1996), "United States Chemical Warfare Policy in World War II: A Captive of Coalition Policy?", *The Journal of Military History* (Society for Military History) **60** (3): 495–511, doi:10.2307/2944522, JSTOR 2944522
- Morton, Desmond; Granatstein, Jack L (1989), *Marching to Armageddon: Canadians and the Great War 1914–1919*, ISBN 0886192099, OCLC 21449019
- Morton, Desmond (1992), *Silent Battle: Canadian Prisoners of War in Germany, 1914–1919*, Toronto: Lester Publishing, ISBN 1895555175, OCLC 29565680
- Mosier, John (2001), "Germany and the Development of Combined Arms Tactics", *Myth of the Great War: How the Germans Won the Battles and How the Americans Saved the Allies*, New York: Harper Collins, ISBN 0060196769
- Muller, Jerry Z (March/April 2008), "Us and Them – The Enduring Power of Ethnic Nationalism" (http://www.foreignaffairs.com/20080301faessay87203/jerry-z-muller/us-and-them.html), *Foreign Affairs* (Council on Foreign Relations), retrieved 30 December 2008
- Neiberg, Michael S (2005), *Fighting the Great War: A Global History*, Cambridge, Mass: Harvard University Press, ISBN 0674016963, OCLC 56592292
- Nicholson, Gerald WL (1962), *Canadian Expeditionary Force, 1914–1919: Official History of the Canadian Army in the First World War* (http://www.censol.ca/research/greatwar/nicholson/index.htm) (1st ed.), Ottawa: Queens Printer and Controller of Stationary, OCLC 2317262
- Northedge, FS (1986), *The League of Nations: Its Life and Times, 1920–1946*, New York: Holmes & Meier, ISBN 0718513169
- Page, Thomas Nelson, *Italy and the World War* (http://net.lib.byu.edu/estu/wwi/comment/Italy/Page04.htm), Brigham Young University, Chapter XI cites "Cf. articles signed XXX in *La Revue de Deux Mondes*, 1 and 15 March 1920"
- Perry, Frederick W (1988), *The Commonwealth armies: manpower and organisation in two world wars*, Manchester University Press, ISBN 9780719025952
- Phillimore, George Grenville; Bellot, Hugh HL (1919), "Treatment of Prisoners of War", *Transactions of the Grotius Society* **5**: 47–64, OCLC 43267276
- Pitt, Barrie (2003), *1918: The Last Act*, Barnsley: Pen and Sword, ISBN 0850529743, OCLC 56468232

- Price, Alfred (1980), *Aircraft versus Submarine: the Evolution of the Anti-submarine Aircraft, 1912 to 1980*, London: Jane's Publishing, ISBN 0710600089, OCLC 10324173 Deals with technical developments, including the first dipping hydrophones
- Prior, Robin (1999), *The First World War*, London: Cassell, ISBN 030435256X
- Raudzens, George (October 1990), "War-Winning Weapons: The Measurement of Technological Determinism in Military History", *The Journal of Military History* (Society for Military History) **54** (4): 403–434, doi:10.2307/1986064, JSTOR 1986064
- Repington, Charles à Court (1920), *The First World War, 1914–1918* (http://www.archive.org/details/firstworldwar19102repiuoft), **2**, London: Constable, ISBN 1113197641
- Rickard, J (5 March 2001), "Erich von Ludendorff, 1865–1937, German General" (http://www.historyofwar.org/articles/people_ludendorff.html), *Military History Encyclopedia on the Web* (HistoryOfWar.org), retrieved 6 February 2008
- Rickard, J (27 August 2007), *The Ludendorff Offensives, 21 March-18 July 1918* (http://www.historyofwar.org/scripts/fluffy/fcp.pl?words=20+July+1918&d=/battles_ludendorff.html)
- Roden, Mike, "The Lost Generation – myth and reality" (http://www.aftermathww1.com/lostgen.asp), *Aftermath – when the boys came home*, retrieved 6 November 2009
- Ross, Stewart Halsey (1996), *Propaganda for War: How the United States was Conditioned to Fight the Great War of 1914–1918*, Jefferson, North Carolina: McFarland, ISBN 0786401117, OCLC 185807544
- Saadi, Abdul-Ilah, *Dreaming of Greater Syria* (http://english.aljazeera.net/focus/arabunity/2008/02/2008525183842614205.html), Al Jazeera English, retrieved 17 November 2009
- Sachar, Howard Morley (1970), *The emergence of the Middle East, 1914–1924*, Allen Lane, ISBN 0713901586, OCLC 153103197
- Safire, William (2008), *Safire's Political Dictionary* (http://books.google.com/?id=jK-0NPoMiYoC), Oxford University Press, ISBN 9780195343342
- Salibi, Kamal Suleiman (1993), "How it all began – A concise history of Lebanon" (http://almashriq.hiof.no/lebanon/900/902/Kamal-Salibi/), *A House of Many Mansions – the history of Lebanon reconsidered*, I.B. Tauris, ISBN 1850430918, OCLC 224705916
- Schindler, J (2003), "Steamrollered in Galicia: The Austro-Hungarian Army and the Brusilov Offensive, 1916", *War in History* **10** (1): 27–59, doi:10.1191/0968344503wh260oa
- Shanafelt, Gary W (1985), *The secret enemy: Austria-Hungary and the German alliance, 1914–1918*, East European Monographs, ISBN 9780880330800
- Shapiro, Fred R; Epstein, Joseph (2006), *The Yale Book of Quotations*, Yale University Press, ISBN 0300107986
- Singh, Jaspal, *History of the Ghadar Movement* (http://www.panjab.org.uk/english/histGPty.html), panjab.org.uk, retrieved 31 October 2007
- Sisemore, James D (2003), *The Russo-Japanese War, Lessons Not Learned* (http://cgsc.cdmhost.com/cdm4/item_viewer.php?CISOROOT=/p4013coll2&CISOPTR=113), U.S. Army Command and General Staff College
- Smele, Jonathan, "War and Revolution in Russia 1914–1921" (http://www.webcitation.org/635RR9gbC), *World Wars in-depth* (BBC), archived from the original (http://www.bbc.co.uk/history/worldwars/wwone/eastern_front_01.shtml) on 9 November 2011, retrieved 12 November 2009
- Speed, Richard B, III (1990), *Prisoners, Diplomats and the Great War: A Study in the Diplomacy of Captivity*, New York: Greenwood Press, ISBN 0313267294, OCLC 20694547
- Stevenson, David (1996), *Armaments and the Coming of War: Europe, 1904–1914*, New York: Oxford University Press, ISBN 0198202083, OCLC 33079190
- Stevenson, David (2004), *Cataclysm: The First World War As Political Tragedy*, New York: Basic Books, pp. 560pp, ISBN 0465081843, OCLC 54001282, major reinterpretation
- Stevenson, David (2005), *The First World War and International Politics*, Oxford: Clarendon, OCLC 248297941
- Gilbert, Martin (1994), *First World War*, Stoddart Publishing, ISBN 9780773728486

- Strachan, Hew (2004), *The First World War: Volume I: To Arms*, New York: Viking, ISBN 0670032956, OCLC 53075929: the major scholarly synthesis. Thorough coverage of 1914
- Strachan, Hew (1998), *The Oxford Illustrated History of the First World War*, New York: Oxford University Press, ISBN 0198206143
- Stumpp, Karl; Weins, Herbert; Smith, Ingeborg W (trans) (1997), *A People on the Move: Germans in Russia and in the Former Soviet Union: 1763–1997* (http://lib.ndsu.nodak.edu/grhc/history_culture/history/people.html), North Dakota State University Libraries
- Swietochowski, Tadeusz (2004), *Russian Azerbaijan, 1905–1920: The Shaping of a National Identity in a Muslim Community*, **42**, Cambridge University Press, ISBN 9780521522458, reviewed at JSTOR 1866737
- Taylor, Alan John Percivale (1963), *The First World War: An Illustrated History*, Hamish Hamilton, OCLC 2054370
- Taylor, Alan John Percivale (1998), *The First World War and its aftermath, 1914–1919*, London: Folio Society, OCLC 49988231
- Taylor, John M (Summer 2007), "Audacious Cruise of the Emden", *The Quarterly Journal of Military History* **19** (4): 38–47, doi:10.1353/jmh.2007.0331 (inactive 2010-07-26), ISSN 0899-3718
- Terraine, John (1963), *Ordeal of Victory*, Philadelphia: Lippincott, pp. 508pp, OCLC 1345833
- Tschanz, David W, *Typhus fever on the Eastern front in World War I* (http://www.entomology.montana.edu/historybug/WWI/TEF.htm), Montana State University, retrieved 12 November 2009
- Tuchman, Barbara Wertheim (1962), *The Guns of August*, New York: Macmillan, OCLC 192333, tells of the opening diplomatic and military manoeuvres
- Tuchman, Barbara Wertheim (1966), *The Zimmerman Telegram* (2nd ed.), New York: Macmillan, ISBN 0026203200, OCLC 233392415
- Tucker, Spencer C (1999), *European Powers in the First World War: An Encyclopedia*, ISBN 081533351X, OCLC 40417794
- Tucker, Spencer C; Roberts, Priscilla Mary (2005), *Encyclopedia of World War I*, Santa Barbara: ABC-Clio, ISBN 1851094202, OCLC 61247250
- Tucker, Spencer C; Wood, Laura Matysek; Murphy, Justin D (1999), *The European powers in the First World War: an encyclopedia*, Taylor & Francis, ISBN 9780815333517
- von der Porten, Edward P (1969), *German Navy in World War II*, New York: T. Y. Crowell, ISBN 021317961X, OCLC 164543865
- Westwell, Ian (2004), *World War I Day by Day*, St. Paul, Minnesota: MBI Publishing, pp. 192pp, ISBN 0760319375, OCLC 57533366
- Wilgus, William John (1931), *Transporting the A. E. F. in Western Europe, 1917–1919*, New York: Columbia University Press, OCLC 1161730
- Willmott, H.P. (2003), *World War I*, New York: Dorling Kindersley, ISBN 0789496275, OCLC 52541937
- Winegard, Timothy, "Here at Vimy: A Retrospective – The 90th Anniversary of the Battle of Vimy Ridge" (http://www.journal.forces.gc.ca/vo8/no2/winegard-eng.asp), *Canadian Military Journal* **8** (2)
- Winter, Denis (1983), *The First of the Few: Fighter Pilots of the First World War*, Penguin, ISBN 9780140052565
- Wohl, Robert (1979), *The Generation of 1914* (3 ed.), Harvard University Press, ISBN 9780674344662
- Zieger, Robert H (2001), *America's Great War: World War I and the American experience*, Lanham, Maryland: Rowman & Littlefield, p. 50, ISBN 0847696456
- "Country Briefings: Israel" (http://www.economist.com/countries/Israel/profile.cfm?folder=History in brief), *The Economist*, 28 July 2005, retrieved 30 December 2008
- Israeli Foreign Ministry, *Ottoman Rule* (http://www.jewishvirtuallibrary.org/jsource/History/Ottoman.html), Jewish Virtual Library, retrieved 30 December 2008

External links

- A multimedia history of World War I (http://www.firstworldwar.com/)
- British Pathé (http://www.britishpathe.com/workspace.php?id=2930&display=list/) Online film archive containing extensive coverage of World War I
- The Heritage of the Great War, Netherlands (http://www.greatwar.nl/)
- The World War I Document Archive (http://wwi.lib.byu.edu/index.php/Main_Page) Wiki, Brigham Young University
- Maps of Europe (http://maps.omniatlas.com/europe/19140905/) covering the history of World War I at omniatlas.com

Animated maps

- An animated map "Europe plunges into war" (http://www.the-map-as-history.com/demos/tome06/)
- An animated map of Europe at the end of the war (http://www.the-map-as-history.com/demos/tome03/)

Hydraulic_brake

The **hydraulic brake** is an arrangement of braking mechanism which uses brake fluid, typically containing ethylene glycol, to transfer pressure from the controlling unit, which is usually near the operator of the vehicle, to the actual brake mechanism, which is usually at or near the wheel of the vehicle.

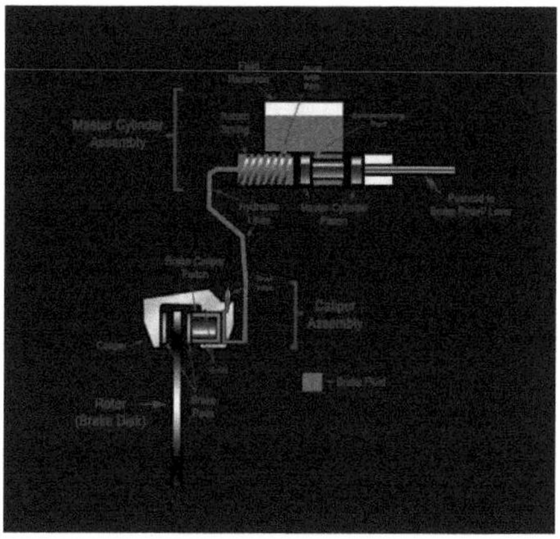

A schematic illustrating the major components of a hydraulic disc brake system

History

In 1918 Malcolm Lougheed (who later changed the spelling of his name to Lockheed) developed a hydraulic brake system.[1]

Construction

The most common arrangement of hydraulic brakes for passenger vehicles, motorcycles, scooters, and mopeds, consists of the following:

- Brake pedal or lever
- A pushrod (also called an *actuating rod*)
- A master cylinder assembly containing a piston assembly (made up of either one or two pistons, a return spring, a series of gaskets/ O-rings and a fluid reservoir)
- Reinforced hydraulic lines
- Brake caliper assembly usually consisting of one or two hollow aluminum or chrome-plated steel pistons (called *caliper pistons*), a set of thermally conductive brake pads and a rotor (also called a *brake disc*) or drum attached to an axle.

The system is usually filled with a glycol-ether based brake fluid (other fluids may also be used).

At one time,the passenger vehicles commonly employed drum brakes on all four wheels. Later, disc brakes were used for the front and drum brakes for the rear. However, because disc brakes have shown a better stopping performance and are therefore generally safer and more effective than drum brakes, four-wheel disc brakes have become increasingly popular, replacing drums on all but the most basic vehicles. Many two-wheel vehicles designs, however, continue to employ a drum brake for the rear wheel.

For simplicity, the braking system described hereafter uses the terminology and configuration for a simple *disc brake*.

System Operation

Within a hydraulic brake system, as the brake pedal is pressed, a pushrod exerts force on the piston(s) in the master cylinder causing fluid from the brake fluid reservoir to flow into a pressure chamber through a compensating port which results in an increase in the pressure of the entire hydraulic system. This forces fluid through the hydraulic lines toward one or more calipers where it acts upon one or two caliper pistons sealed by one or more seated O-rings which prevent the escape of any fluid from around the piston.

The brake caliper pistons then apply force to the brake pads. This causes them to be pushed against the spinning rotor, and the friction between the pads and the rotor causes a braking torque to be generated, slowing the vehicle. Heat generated from this friction is either dissipated through vents and channels in the rotor or conducted through the pads themselves which are made of specialized heat-tolerant materials such as kevlar or sintered glass.

Subsequent release of the brake pedal/ lever allows spring(s) to return the master piston(s) back into position. This relieves the hydraulic pressure on the caliper allowing the brake piston in the caliper assembly to slide back into its housing and the brake pads to release the rotor. The hydraulic braking system is designed as a closed system: unless there is a leak within the system, none of the brake fluid enters or leaves it, nor does it get consumed through use.

An example of a hydraulic brake system

Hydraulic brakes transfer energy to stop an object, normally a rotating axle. In a very simple brake system, with just two cylinders and a disc brake the cylinders could be connected via tubes, with a piston inside the cylinders. The cylinders and tubes are filled with incompressible oil. The two cylinders have the same volume, but different diameters, therefore a different surface areas. The one with the smallest diameter is called the master cylinder. The spinning disc brake, will be placed down at the piston with the larger surface area. Let us say that the diameter of the master cylinder is half the diameter of the slave cylinder, so the master cylinder has a surface area which is four times smaller. Now, if the piston in the master cylinder is pushed down 40 mm, with 10 N of force, the slave piston will then move 10 mm, with a force of 40 N.

This force can be further multiplied by adding a lever connected between the master piston, a pedal, and a pivot point. If the distance from the pedal to the pivot is three times the distance from the pivot to the connected piston, then it multiplies the pedal force with a factor of 3, when pushing down on the pedal. Now, if we push down on the pedal 120 mm with 10 N of force, 30 N will then be applied to the master piston, and the slave piston will then push 10 mm against the brake pad, with a total force of 120 N.

Component specifics

(For typical light duty automotive braking systems)

The **brake pedal** is a simple lever. One end is attached to the framework of the vehicle, a pushrod extends from a point along its length, and the foot pad is at the other end of the lever. The rod either extends to the master cylinder (manual brakes) or to the vacuum booster (power brakes).

In a four-wheel car, the master cylinder is divided internally into two sections, each of which pressurizes a separate hydraulic circuit. Each section supplies pressure to one circuit. Passenger vehicles typically have either a front/rear split brake system or a diagonal **split brake system** (the master cylinder in a motorcycle or scooter may only pressurize a single unit, which will be the front brake).

A front/rear split system uses one master cylinder section to pressurize the front caliper pistons and the other section to pressurize the rear caliper pistons. A split circuit braking system is now required by law in most countries for safety reasons; if one circuit fails, the other circuit can stop the vehicle.

Diagonal split systems were used initially on American Motors automobiles in the 1967 production year. The right front and left rear are served by one actuating piston while the left front and the right rear are served, exclusively, by a second actuating piston (both pistons pressurize their respective coupled lines from a single foot pedal). If either circuit fails, the other, with at least one front wheel braking (the front brakes provide most of the speed reduction) remains intact to stop the mechanically-damaged vehicle. Just before 1970, diagonally split systems had become universal for automobiles sold in the United States.

The diameter and length of the master cylinder has a significant effect on the performance of the brake system. A larger diameter master cylinder delivers more hydraulic fluid to the caliper pistons, yet requires more brake pedal force and less brake pedal stroke to achieve a given deceleration. A smaller diameter master cylinder has the opposite effect.

A master cylinder may also use differing diameters between the two sections to allow for increased fluid volume to one set of caliper pistons or the other.

Power brakes

The **vacuum booster** or **vacuum servo** is used in most modern hydraulic brake systems which contain four wheels. The vacuum booster is attached between the master cylinder and the brake pedal and multiplies the braking force applied by the driver. These units consist of a hollow housing with a movable rubber diaphragm across the center, creating two chambers. When attached to the low-pressure portion of the throttle body or intake manifold of the engine, the pressure in both chambers of the unit is lowered. The equilibrium created by the low pressure in both chambers keeps the diaphragm from moving until the brake pedal is depressed. A return spring keeps the diaphragm in the starting position until the brake pedal is applied. When the brake pedal is applied, the movement opens an air valve which lets in atmospheric pressure air to one chamber of the booster. Since the pressure becomes higher in one chamber, the diaphragm moves toward the lower pressure chamber with a force created by the area of the diaphragm and the differential pressure. This force, in addition to the driver's foot force, pushes on the master cylinder piston. A relatively small diameter booster unit is required; for a very conservative 50% manifold vacuum, an assisting force of about 1500 N (200n) is produced by a 20 cm diaphragm with an area of 0.03 square meters. The diaphragm will stop moving when the forces on both sides of the chamber reach equilibrium. This can be caused by either the air valve closing (due to the pedal apply stopping) or if "run out" is reached. Run out occurs when the pressure in one chamber reaches atmospheric pressure and no additional force can be generated by the now stagnant differential pressure. After the run out point is reached, only the driver's foot force can be used to further apply the master cylinder piston.

The fluid pressure from the master cylinder travels through a pair of steel brake tubes to a **pressure differential valve**, sometimes referred to as a "brake failure valve", which performs two functions: it equalizes pressure between the two systems, and it provides a warning if one system loses pressure. The pressure differential valve has two chambers (to which the hydraulic lines attach) with a piston between them. When the pressure in either line is balanced, the piston does not move. If the pressure on one side is lost, the pressure from the other side moves the piston. When the piston makes contact with a simple electrical probe in the center of the unit, a circuit is completed, and the operator is warned of a failure in the brake system.

From the pressure differential valve, brake tubing carries the pressure to the brake units at the wheels. Since the wheels do not maintain a fixed relation to the automobile, it is necessary to use hydraulic brake hose from the end of the steel line at the vehicle frame to the caliper at the wheel. Allowing steel brake tubing to flex invites metal fatigue and, ultimately, brake failure. A common upgrade is to replace the standard rubber hoses with a set which are externally reinforced with braided stainless-steel wires; these have negligible expansion under pressure and can give a firmer feel to the brake pedal with less pedal travel for a given braking effort.

Special considerations

Air brake systems are bulky, and require air compressors and reservoir tanks. Hydraulic systems are smaller and less expensive.

Hydraulic fluid must be non-compressible. Unlike air brakes, where a valve is opened and air flows into the lines and brake chambers until the pressure rises sufficiently, hydraulic systems rely on a single stroke of a piston to force fluid through the system. If any vapor is introduced into the system it will compress, and the pressure may not rise sufficiently to actuate the brakes.

Hydraulic braking systems are sometimes subjected to high temperatures during operation, such as when descending steep grades. For this reason, hydraulic fluid must resist vaporization at high temperatures.

Water vaporizes easily with heat and can corrode the metal parts of the system. If it gets into the brake lines, it can degrade brake performance dramatically. This is why light oils are often used as hydraulic fluids. Oil displaces water, protects plastic parts against corrosion, and can tolerate much higher temperatures before vaporizing.

"Brake fade" is a condition caused by overheating in which braking effectiveness reduces, and may be lost. It may occur for many reasons. The pads which engage the rotating part may become overheated and "glaze over", becoming so smooth and hard that they cannot grip sufficiently to slow the vehicle. Also, vaporization of the hydraulic fluid under temperature extremes or thermal distortion may cause the linings to change their shape and engage less surface area of the rotating part. Thermal distortion may also cause permanent changes in the shape of the metal components, resulting in a reduction in braking capability that requires replacement of the affected parts.

See also

- Hydraulics
- Hydraulic circuit
- Fuse (hydraulic)
- Bicycle brake systems
- Vehicle brake
- Brake bleeding
- Brake-by-wire

External links

- Nice, Karim. "How Brakes Work" [2]. How Stuff Works. Retrieved 18 June 2010.
- "Hydraulic Brakes" [3]. Integrated Publishing. Retrieved 18 June 2010.

Patents

- US 2746575 [4]. Kinchin 1956-05-22
- US 2591793 [5]. Dubois 1952-04-08
- US 2544849 [6]. Martin 1951-03-13
- US 2485032 [7]. Bryant 1949-10-08
- US 2466990 [8]. Johnson Wade C, Trishman Harry A, Stratton Edgar H. 1949-04-12
- US 2416091 [9]. Fitch 1947-02-12
- US 2405219 [10]. Lambert Homer T. 1946-08-06
- US 2375855 [11]. Lambert Homer T. 1945-05-15
- US 2366093 [12]. Forbes Joseph A. 1944-12-26
- US 2140752 [13]. La Brie 1938-12-20
- US 2084216 [14]. Poage Robert A. and Poage Marlin Z. 1937-06-15
- US 2028488 [15]. Avery William Leicester 1936-02-21
- US 1959049 [16]. Buus Niels Peter Valdemar 1934-05-15
- US 1954534 [17]. Norton Raymond J 1934-04-10
- US 1721370 [18]. Boughton Edward Bishop 1929-07-16
- DE 695921 [19]. Borgwar Carl Friedrich Wilhelm 1940-09-06
- GB 377478 [20]. Hall Frederick Harold 1932-07-28
- GB 365069 [21]. Rubury John Meredith 1932-01-06

References

[1] Csere, Csaba (January 1988), "10 Best Engineering Breakthroughs", *Car and Driver* **33** (7), p. 61.

[2] http://auto.howstuffworks.com/auto-parts/brakes/brake-types/brake1.htm

[3] http://www.tpub.com/fluid/ch1j.htm

[4] http://worldwide.espacenet.com/textdoc?DB=EPODOC&IDX=US2746575

[5] http://worldwide.espacenet.com/textdoc?DB=EPODOC&IDX=US2591793

[6] http://worldwide.espacenet.com/textdoc?DB=EPODOC&IDX=US2544849

[7] http://worldwide.espacenet.com/textdoc?DB=EPODOC&IDX=US2485032

[8] http://worldwide.espacenet.com/textdoc?DB=EPODOC&IDX=US2466990

[9] http://worldwide.espacenet.com/textdoc?DB=EPODOC&IDX=US2416091

[10] http://worldwide.espacenet.com/textdoc?DB=EPODOC&IDX=US2405219

[11] http://worldwide.espacenet.com/textdoc?DB=EPODOC&IDX=US2375855

[12] http://worldwide.espacenet.com/textdoc?DB=EPODOC&IDX=US2366093

[13] http://worldwide.espacenet.com/textdoc?DB=EPODOC&IDX=US2140752

[14] http://worldwide.espacenet.com/textdoc?DB=EPODOC&IDX=US2084216

[15] http://worldwide.espacenet.com/textdoc?DB=EPODOC&IDX=US2028488

[16] http://worldwide.espacenet.com/textdoc?DB=EPODOC&IDX=US1959049

[17] http://worldwide.espacenet.com/textdoc?DB=EPODOC&IDX=US1954534

[18] http://worldwide.espacenet.com/textdoc?DB=EPODOC&IDX=US1721370

[19] http://worldwide.espacenet.com/textdoc?DB=EPODOC&IDX=DE695921

[20] http://worldwide.espacenet.com/textdoc?DB=EPODOC&IDX=GB377478

[21] http://worldwide.espacenet.com/textdoc?DB=EPODOC&IDX=GB365069

- Erjavec, Jack (2004). *Automotive Technology: A Systems Approach*, Delmar Cengage Learning. ISBN 1401848311
- Allan and Malcolm Loughead (Lockheed) Their Early Lives in the Santa Cruz Mountains including the invention of the hydraulic brake. (http://www.wmsb.net/loughead-lockheed.html)

Tours

Tours	

Town hall and Place Jean Jaurès	
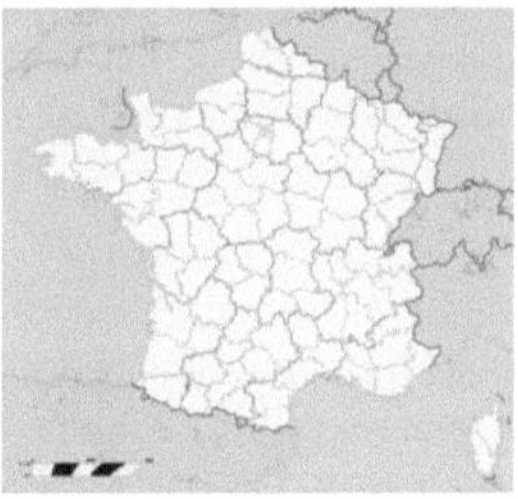	
Tours	
Administration	
Country	France
Region	
Department	
Arrondissement	Tours
Mayor	Jean Germain (2008–2014)
Statistics	
Elevation	44–109 m (**unknown operator: u'strong'unknown operator: u'strong'unknown operator: u'strong' unknown operator: u'strong'**)
Land area[1]	34.36 km^2 (**unknown operator: u'strong'** sq mi)
Population[2]	135218 (2009)
- Density	3935 /km^2 (**unknown operator: u'strong'** /sq mi)
INSEE/Postal code	37261 [1]/ 37000, 37100, 37200
[1] French Land Register data, which excludes lakes, ponds, glaciers > 1 km^2 (0.386 sq mi or 247 acres) and river estuaries.	

[2] *Population without double counting*: residents of multiple communes (e.g., students and military personnel) only counted once.

Tours (pronounced: [tuʁ]) is a city in central France, the capital of the Indre-et-Loire department.

It stands on the lower reaches of the river Loire, between Orléans and the Atlantic coast. Touraine, the region around Tours, is known for its wines, the alleged perfection (as perceived by some speakers) of its local spoken French, and the Battle of Tours in 732. It is also the site of the Paris–Tours road bicycle race. Tours is the largest city in the Centre region of France, although it is not the regional capital, which is the region's second-largest city, Orléans. In 2006, the city itself had 142,000 inhabitants and the metropolitan area had 306,974.

History

In Gallic times the city was important as a crossing point of the Loire. Becoming part of the Roman Empire during the first century AD, the city was named "Caesarodunum" ("hill of Caesar"). The name evolved in the 4th century when the original Gallic name, Turones, became first "Civitas Turonum" then "Tours". It was at this time that the amphitheatre of Tours, one of the five largest in the Empire, was built. Tours became the metropolis of the Roman province of Lugdunum towards 380–388, dominating the Loire Valley, Maine and Brittany. One of the outstanding figures of the history of the city was Saint Martin, second bishop who shared his coat with a naked beggar in Amiens. This incident and the importance of Martin in the medieval Christian West made Tours, and its position on the route of pilgrimage to Santiago de Compostela, a major centre during the Middle Ages.

Middle Ages

In the 6th century Gregory of Tours, author of the *Ten Books of History*, made his mark on the town by restoring the cathedral destroyed by a fire in 561. Saint Martin's monastery benefited from its inception, at the very start of the 6th century from patronage and support from the Frankish king, Clovis, which increased considerably the influence of the saint, the abbey and the city in Gaul. In the 9th century, Tours was at the heart of the Carolingian Rebirth, in particular because of Alcuin abbot of Marmoutier.

In 732 AD, Abdul Rahman Al Ghafiqi and a large army of Muslim horsemen from Al-Andalus advanced 500 km deep into France, and were stopped at Tours by Charles Martel and his infantry igniting the Battle of Tours. The outcome was defeat for the Muslims, preventing France from Islamic conquest. In 845, Tours repulsed the first attack of the Viking chief Hasting (Haesten). In 850, the Vikings settled at the mouths of the Seine and the Loire. Still led by Hasting, they went up the Loire again in 852 and sacked Angers, Tours and the abbey of Marmoutier.

During the Middle Ages, Tours consisted of two juxtaposed and competing centres. The "City" in the east, successor of the late Roman 'castrum', was composed of the archiepiscopal establishment (the cathedral and palace of the archbishops) and of the castle of Tours, seat of the authority of the Counts of Tours (later Counts of Anjou) and of the King of France. In the west, the "new city" structured around the Abbey of Saint Martin was freed from the control of the City during the 10th century (an enclosure was built towards 918) and became "Châteauneuf". This space, organized between Saint Martin and the Loire, became the economic centre of Tours. Between these two centres remained Varenne, vineyards and fields, little occupied except for the Abbaye Saint-Julien established on the banks of the Loire. The two centres were linked during the 14th century. Tours is a good example of a medieval double city.

Tours became the capital of the county of Tours or Touraine, territory bitterly disputed between the counts of Blois and Anjou – the latter were victorious in the 9th century. It was the capital of France at the time of Louis XI, who had settled in the castle of Montils (today the castle of Plessis in La Riche, western suburbs of Tours), Tours and Touraine remained until the 16th century a permanent residence of the kings and court. The rebirth gave Tours and

Touraine many private mansions and castles, joined together to some extent under the generic name of the Chateaux of the Loire. It is also at the time of Louis XI that the silk industry was introduced – despite difficulties, the industry still survives to this day.

16th–18th centuries

Charles IX passed through the city at the time of his royal tour of France between 1564 and 1566, accompanied by the Court and various noblemen: his brother the Duke of Anjou, Henri de Navarre, the cardinals of Bourbon and Lorraine. At this time, the Catholics returned to power in Angers: the intendant assumed the right to nominate the aldermen. The Massacre of Saint-Barthelemy was not repeated at Tours. The Protestants were imprisoned by the aldermen – a measure which prevented their extermination. The permanent return of the Court to Paris and then Versailles marked the beginning of a slow but permanent decline. Guillaume the Metayer (1763–1798), known as Rochambeau, the well known counter-revolutionary chief of Mayenne, was shot there on Thermidor 8, year VI.

19th–20th centuries

However, it was the arrival of the railway in the 19th century which saved the city by making it an important nodal point. The main railway station is known as Tours-Saint-Pierre-des-Corps. At that time, Tours was expanding towards the south into a district known as the Prébendes. The importance of the city as a centre of communications contributed to its revival and, as the 20th century progressed, Tours became a dynamic conurbation, economically oriented towards the service sector.

First World War

The city was greatly affected by the First World War. A force of 25,000 American soldiers arrived in 1917, setting up textile factories for the manufacture of uniforms, repair shops for military equipment, munitions dumps, an army post office and an American military hospital at Augustins. Thus Tours became a garrison town with a resident general staff. The American presence is remembered today by the Woodrow Wilson bridge over the Loire, which was officially opened in July 1918 and bears the name of the man who was President of the USA from 1912 to 1920. Three American air force squadrons, including the 492nd, were based at the Parçay-Meslay airfield, their personnel playing an active part in the life of the city. Americans paraded at funerals and award ceremonies for the Croix de Guerre; they also took part in festivals and their YMCA organised shows for the troops. Some men married girls from Tours.

Tours Cathedral: 15th century Flamboyant Gothic west front with Renaissance pinnacles, completed 1547.

Inter-war years

In 1920, the city was host to the Congress of Tours, which saw the creation of the French Communist Party.

Second World War

Tours was also marked by the Second World War. In 1940, the city suffered massive destruction and for four years it was a city of military camps and fortifications. From 10–13 June 1940, Tours was the temporary seat of the French government before its move to Bordeaux. German incendiary bombs caused a huge fire which blazed out of control from 20–22 June and destroyed part of the city centre. Some architectural masterpieces of the 16th and 17th centuries were lost, as was the monumental entry to the city. The Wilson Bridge (known locally as the 'stone bridge'), carried a water main which supplied the city; the bridge was dynamited to slow the progress of the German advance. With the water main severed and unable to extinguish the inferno, the inhabitants had no option but to flee to safety. More heavy air raids devastated the area around the railway station in 1944 causing several hundred deaths.

Post-war developments

A plan for the rebuilding of the downtown area drawn up by the local architect Camille Lefèvre was adopted even before the end of the war. The plan was for 20 small quadrangular blocks of housing to be arranged around the main road (la rue Nationale), which was widened. This regular layout attempted to echo, yet simplify, the 18th century architecture. Pierre Patout succeeded Lefèvre as the architect in charge of rebuilding in 1945. At one time there was talk of demolishing the southern side of the rue Nationale in order to make it in keeping with the new development.

The recent history of Tours is marked by the personality of Jean Royer, who was Mayor for 36 years and helped to save the old town from demolition by establishing one of the first Conservation Areas. This example of conservation policy would later inspire the Malraux Law for the safeguarding of historic city centres. In the 1970s, Jean Royer also extended the city to the south by diverting the course of the River Cher to create the districts of Rives du Cher and des Fontaines; at the time, this was one of the largest urban developments in Europe. In 1970, the François-Rabelais university was founded; this is centred on the bank of the Loire in the downtown area, and not – as it was then the current practice – in a campus in the suburbs. The latter solution was also chosen by the twin university of Orleans. Royer's long term as Mayor was, however, not without controversy, as exemplified by the construction of the practical – but aesthetically unattractive – motorway which runs along the bed of a former canal just 1500 metres from the cathedral. Another bone of contention was the original Vinci Congress Centre by Jean Nouvel. This project incurred debts although it did, at least, make Tours one of France's principal conference centres.

Jean Germain became Mayor in 1995 and made debt reduction his priority. Ten years later, his economic management is regarded as much wiser than that of his predecessor, the financial standing of the city having returned to a stability. However, the achievements of Jean Germain are criticised by the municipal opposition for a lack of ambition: no large building projects comparable with those of Jean Royer have been instituted under his double mandate. This position is disputed by those in power, who affirm their policy of concentrating on the quality of life, as evidenced by urban restoration, the development of public transport and cultural activities.

Population

Historical populations		
Year	**Pop.**	**±%**
1793	21000	—
1800	20240	−3.6%
1806	21703	+7.2%
1821	21928	+1.0%
1831	23235	+6.0%
1836	26669	+14.8%
1841	30072	+12.8%
1846	30766	+2.3%
1851	33530	+9.0%
1856	38055	+13.5%
1861	41061	+7.9%
1866	42450	+3.4%
1872	43368	+2.2%
1876	48325	+11.4%
1881	52209	+8.0%
1886	59585	+14.1%
1891	60335	+1.3%
1896	63267	+4.9%
1901	64695	+2.3%
1906	67601	+4.5%
1911	73398	+8.6%
1921	75096	+2.3%
1926	77192	+2.8%
1931	78585	+1.8%
1936	83753	+6.6%
1946	80044	−4.4%
1954	83618	+4.5%
1962	92944	+11.2%
1968	128120	+37.8%
1975	140686	+9.8%
1982	132209	−6.0%
1990	129509	−2.0%
1999	132677	+2.4%
2006	136942	+3.2%
2009	135218	−1.3%

Sights

Place Jean Jaurès.

Tours Cathedral

The cathedral of Tours, dedicated to Saint Gatien, its canonized first bishop, was begun about 1170 to replace the cathedral that was burnt out in 1166, during the dispute between Louis VII of France and Henry II of England. The lowermost stages of the western towers (*illustration, above left*) belong to the 12th century, but the rest of the west end is in the profusely detailed Flamboyant Gothic of the 15th century, completed just as the Renaissance was affecting the patrons who planned the châteaux of Touraine. These towers were being constructed at the same time as, for example, the Château de Chenonceau.

St Gatien Cathedral, from Rue Lavoisier, just north of the Rue Colbert intersection.

When the 15th century illuminator Jean Fouquet was set the task of illuminating Josephus's *Jewish Antiquities*, his depiction of Solomon's Temple was modeled after the nearly-complete cathedral of Tours. The atmosphere of the Gothic cathedral close permeates Honoré de Balzac's dark short novel of jealousy and provincial intrigues, *Le Curé de Tours* (*The Curate of Tours*) and his medieval story *Maitre Cornélius* opens within the cathedral itself.

Other points of interest

- Jardin botanique de Tours, the municipal botanical garden
- Musée des Beaux-Arts de Tours
- Hôtel Goüin

Pont Wilson.

Language

The inhabitants of Tours (*Les Tourangeaux*) are renowned for speaking the "purest" form of French in the entire country. The pronunciation of Touraine is traditionally regarded as the most standard pronunciation of the French language, supposedly devoid of any perceived accent (unlike that of most other regions of France, including Paris). Gregory of Tours wrote in the 6th century that some people in this area could still speak Gaulish.

City

The city of Tours has a population of 140,000 and is called "Le Jardin de la France" ("The Garden of France"). There are several parks located within the city. Tours is located between two rivers, the Loire to the north and the Cher to the south. The buildings of Tours are white with blue slate (called *Ardoise*) roofs; this style is common in the north of France, while most buildings in the south of France have terracotta roofs.

Tours is famous for its original medieval district, called *le Vieux Tours*. Unique to the Old City are its preserved half-timbered buildings and *la Place Plumereau*, a square with busy pubs and restaurants, whose open-air tables fill the centre of the square. The Boulevard Beranger crosses the Rue Nationale at the Place Jean-Jaures and is the location of weekly markets and fairs.

Tours is famous for its many bridges crossing the river Loire. One of them, the Pont Wilson, collapsed in 1978, but was rebuilt just like it was before.

Near the cathedral, in the garden of the ancient Palais des Archevêques (now *Musée des Beaux-Arts*), is a huge cedar tree planted by Napoleon. The garden also has in an alcove a stuffed elephant, Fritz. He escaped from the Barnum and Bailey circus during their stay in Tours in 1902. He went mad and had to be shot down, but the city paid to honor him, and he was stuffed as a result.

Tours is home to François Rabelais University, the site of one of the most important choral competitions, called *Florilège Vocal de Tours* International Choir Competition, and is a member city of the European Grand Prix for Choral Singing.

Transport

Tram model, design by the french agency *RCP Design Global*

Today, with its extensive rail (including TGV) and autoroute links to the rest of the country, Tours is a jumping-off point for tourist visits to the Loire Valley and the royal chateaux.

Tours is on one of the main lines of the TGV. It is possible to travel to the west coast at Bordeaux in two and a half hours, to the Mediterranean coast via Avignon and from there to Spain and Barcelona, or to Lyon, Strasbourg and Lille. It takes less than one hour by train from Tours to Paris by TGV and one hour and a half to Charles de Gaulle airport. Tours has two main stations: the central station Gare de Tours, and Gare de Saint-Pierre-des-Corps, just outside the centre, the station used by trains that do not terminate in Tours.

Tours Loire Valley Airport connects the Loire Valley to London Stansted Airport and Dublin. These links are provided by the Irish airline Ryanair. National connections to Figari on Corsica, Porto and Marseille are also available during the summer.

Tours does not have a metro rail system; instead there is a bus service, the main central stop being *Jean Jaures,* which is next to the Hôtel de Ville, and *rue Nationale*, the high street of Tours. A tram network is under construction; completion is expected in September 2013, and 21 Citadis trams have been ordered from Alstom designed by RCP Design Global.[2]

Catholics from Tours

Tours is a special place for Catholics who follow the devotion to the Holy Face of Jesus and the adoration of the Blessed Sacrament. It was in Tours in 1843 that a Carmelite nun, Sister Marie of St Peter reported a vision which started the devotion to the Holy Face of Jesus, in reparation for the many insults Christ suffered in His Passion. The *Golden Arrow Prayer* was first made public by her in Tours.

Venerable Leo Dupont, *Holy Man of Tours*

The Venerable Leo Dupont also known as The Holy Man of Tours lived in Tours at about the same time. In 1849 he started the nightly adoration of the Blessed Sacrament in Tours, from where it spread within France. Upon hearing of Sister Marie of St Peter's reported visions, he started to burn a vigil lamp continuously before a picture of the Holy Face of Jesus and helped spread the devotion within France. The devotion was eventually approved by Pope Pius XII in 1958 and he formally declared the Feast of the Holy Face of Jesus as Shrove Tuesday (the Tuesday before Ash Wednesday) for all Roman Catholics.[3] The Oratory of the Holy Face on Rue St. Etienne in Tours receives many pilgrims every year.

Tours was the site of the episcopal activity of St. Martin of Tours and has further Christian connotations in that the pivotal Battle of Tours in 732 is often considered the very first decisive victory over the invading Islamic forces, turning the tide against them. The battle also helped lay the foundations of the Carolingian Empire[4]

Personalities

Tours was the birthplace of:

- Honoré de Balzac (1799–1850), novelist
- Marie of St Peter (1816–1848), mystic carmelite nun
- Berengarius of Tours (999–1088), theologian
- Bernard of Tours (fl. 1147, d. before 1178), philosopher and poet
- Ernest Goüin (1815–1885), french engineer
- Yves Bonnefoy (born 1923), poet
- Abraham Bosse (1604–1676), artist
- Georges Courteline (1858–1929), dramatist and novelist
- Emile Delahaye (1843–1905), automobile pioneer
- Philippe Néricault Destouches (1680–1754), dramatist
- Jean Fouquet (1420–1481), painter
- Gabriel Lamé (1795–1870), mathematician
- Josselin Ouanna tennis player
- Laurent Petitguillaume (1960), radio and television host
- Nâdiya (1973), singer
- Harry Roselmack television presenter
- Ludovic Roy footballer
- Philippe de Trobriand (1816–1897), author, American military officer
- Louise de la Vallière (1644–1710), courtesan
- Zaz (born 1980), singer
- Luc Ducalcon (born 1984), rugby union player

International relations

Tours is twinned with:

- Mülheim an der Ruhr, Germany, since 1962
- Segovia, Spain, since 1972
- Parma, Italy, since 1976
- Luoyang, People's Republic of China, since 1982
- Springfield, Missouri, USA, since 1984
- Trois-Rivières, Canada, since 1987
- Takamatsu, Japan, since 1988
- Braşov, Romania, since 1990
- Minneapolis, Minnesota USA, since 1991

Jardin de la Préfecture (central park).

Giant Cedar tree.

Hôtel Goüin.

Place Plumereau, Medieval buildings.

Place Plumereau.

Slate rooves of Tours.

Looking towards central Tours from the north bank of the River Loire, adjacent to the Pont Mirabeau.

See also

- Bishop of Tours
- University of Tours
- Tours FC – a soccer club based in the town
- Communes of the Indre-et-Loire department

References

- INSEE commune file [5]

[1] http://recensement.insee.fr/searchResults.action?zoneSearchField=&codeZone=37261-COM
[2] "Tours selects Citadis and APS" (http://www.railwaygazette.com/nc/news/single-view/view/tours-selects-citadis-and-aps.html). *Railway Gazette International* (London). 14 September 2010. . Retrieved 15 September 2010.
[3] Dorothy Scallan. "The Holy Man of Tours." (1990) ISBN 0-89555-390-2
[4] Davis, Paul K. (1999) "100 Decisive Battles From Ancient Times to the Present" ISBN 0-19-514366-3
[5] http://www.insee.fr/fr/ppp/bases-de-donnees/recensement/populations-legales/commune.asp?depcom=37261&annee=2009

External links

- Official website (http://www.tours.fr/) (French)
- Tours on French version of Wikipedia
- Tours and its agglomeration on video (http://www.tours-agglo.fr/media/media19058.wmv) (from www.toursmetropole.fr)
- Architecture of Tours (http://france.archiseek.com/indre_et_loire/tours/index.html)
- François Rabelais University, Tours (http://www.univ-tours.fr)
- Satellite picture by Google Maps (http://maps.google.com/maps?ll=47.394333,0.693512&spn=0.112383,0.234180&t=k&hl=en)

Article Sources and Contributors

Rolland-Pilain *Source*: http://en.wikipedia.org/w/index.php?title=Rolland-Pilain *Contributors*: KaptenKrater

San_Sebastián_Grand_Prix *Source*: http://en.wikipedia.org/w/index.php?title=San_Sebasti%C3%A1n_Grand_Prix *Contributors*: Andrwsc, DH85868993, JaGa, John Anderson, KaptenKrater, LilHelpa, NaBUru38, Noym, Sporti

24_Hours_of_Le_Mans *Source*: http://en.wikipedia.org/w/index.php?title=24_Hours_of_Le_Mans *Contributors*: 25or6to4, 72Dino, A Train, AB-me, AMCKen, Abigail-II, Adamml13, Afri, After Midnight, Ahkond, Ahoerstemeier, Ajk, Alai, Aldie, AlexJ, Alpha Centaury, Andreasu, Andy29, Arem250, Art LaPella, Ashley Pomeroy, Atomic7732, Austinknob, AyrtonSenna, B.d.mills, BD2412, Ballista, Banaticus, Ben Ben, BesigedB, Betacommand, BigOg67, BillCook, Bjmullan, Bobblewik, Boojum, Brianhenke, Brianjd, Brighterorange, Brossow, Brutaldeluxe, Bryan Derksen, BryanHolland, Captain Kirk, Carnildo, Chmod007, Chr.K., Chris Ssk, ChristalPalace, Cjeffreys, Cliché Online, Cmdrjameson, Conversion script, CrazyVas, CygnusPius, D, Daguerio, Darklilac, David.Monniaux, Davidstrauss, Ddebree, Delroyb, Deon Steyn, DerHexer, Derek Ross, Desertpad, Digangi96, Djr xi, Dodiad, Donnie Park, Dpdr, Dr.tux, Drdisque, Drmotley, Ducklingmcfc, Dysprosia, E36GUY, ENeville, EagleFan, Echuck215, EeepEeep, El Paulio, Elideb, Emersoni, Epbr123, Epiales, Epolk, Eric119, Ericd, Ericoides, Espsko, EurekaLott, Exxy, Ezn, FCYTravis, Falcadore, Firsfron, Flapdragon, Fox1, Francs2000, Fratrep, Fred Bradstadt, Fredrik, FstrthnU, Fuzzyworcester, Fxhomie, GMacE, GVOLTT, Gaius Cornelius, Gareth E Kegg, Gcm, Gentgeen, Gfoley4, Giraffedata, Gmingst, Gogo Dodo, GregorB, Grunt, Gzuckier, Hailey C. Shannon, Hcheney, Hektikhek, HenkvD, Hephaestos, Hezzy, Hnfr, Hooperbloob, Hotlorp, HunterX, Hvn0413, Ian Dalziel, Ian Pitchford, Icelan, Ida Shaw, Iliyastefanov, Impreziv, Isfisk, Isnow, J-boogie, JLeclerc, JaGa, Jacob Poon, Jacques Delson, Jagfan50, JamesBWatson, Jamesb01, Jamesdennison, Jamesontai, Janothird, Jaraalbe, JayC, Jedikaiti, Jeff Muscato, Jejking, Jeronimo, Jgr64, JillandJack, Jivecat, Jllm06, Joe Canuck, Joel7687, Joema, John Nevard, John of Reading, JohnnyBGood, Johnwill45, Julianhall, Jupix, JustinH, K1Bond007, K3rb, Kaare, Kahastok, Katefan0, Kehrbykid, Kingturtle, Krótki, L33tminion, Lanway, Lethalox, LiamE, Liftarn, Lightmouse, LilHelpa, Lindberg, Linuxmatt, Lipatden, Lipmingarolnick, Llywrch, Loganberry, Loopy, LuNatic, LuxNevada, M-le-mot-dit, Malleus Fatuorum, Malo0178, Mandarax, Maracana, Marc Lacoste, Mark83, Markj 87, Markomagare, MarnetteD, Marsman4444, Martpol, Matthead, Maxwellversion2, McSly, Mentifisto, Mervyn, Metal Militia, Mfraioli, Mgtc77, MickMacNee, Mike65535, Minesweeper, Mintguy, MisfitToys, Missmarple, Miterdale, Mooquackwooftweetmeow, Morio, Moriori, Morven, Mr Gress3000, Mu5ti, Muad, Mulsanne253, Munfis, N328KF, NaBUru38, Nascarking, Natalie Erin, NightCrawler, Nknight, Nommonomanac, Noypi380, Officially Mr X, Ohconfucius, Okumaguru, Old Guard, Olivier, Osomec, PIRO700, Paul590, Pburka, Pc13, Per Abrahamsen, Peter Horn, Philip Trueman, Phydaux, Pinizicc65, Pmsyyz, Pok148, Pol098, Pollinator, Pontificalibus, Pruneau, Psbsub, Pyrope, Quadell, R. sparts, RAE, RJFJR, Ranger2000, Recury, Redvers, RexNL, Rhe br, Rich Farmbrough, Richard n, Rjwilmsi, Rmhermen, Rmpeach5, Robertjohnsonrj, Rogerd, RpM Mad Monk, RyJones, SGBailey, SM, SMC, Sabre ball, Sam Hocevar, Samstayton, Saschaporsche, SaviorSix, SchfiftyThree, SchrodingersMutt, Schwimmnudel07, Scjessey, Scott Paeth, SeeSchloss, Segye, Serrin, Sesesq, Sfoskett, Shanes, Shimgray, Simontemplar 007, SkyLined, Snowdog, Some jerk on the Internet, Soquette, Sporti, Spyder Monkey, StaticGull, Stephanbim, Stevemayhew, StormIT, Suwa, TGCP, TJ Spyke, TJSwoboda, Tabletop, TacoJim, Teutonic Tamer, The359, TheYmode, Thumperward, Timc, Timppis, Tomvmax, Trekphiler, Tresiden, Trevor MacInnis, Triikan, Troels Nybo, Typ932, Ulric1313, Ultimatum, Uncle Scrooge, Valentinian, Wayne Slam, WikHead, WikiPino, Wikigi, Will Pittenger, William Leadford, Willirennen, Wizzy, WojPob, WrcF1, Yann, Yuriybrisk, Zzyzx11, [212], 551 anonymous edits

Albert_Guyot *Source*: http://en.wikipedia.org/w/index.php?title=Albert_Guyot *Contributors*: Byetime, Chief tin cloud, Colonies Chris, Drdisque, Mu, Mycroft.Holmes, Npeters22, Saxbryn, Sporti, Waacstats

Grand_Prix_motor_racing *Source*: http://en.wikipedia.org/w/index.php?title=Grand_Prix_motor_racing *Contributors*: 4u1e, Addihockey10 (automated), Adrian M. H., Ame A1GP, Andrwsc, Angr, Annelid, Armbrust, Autair, BlackJack, Brossow, Bryan Derksen, C. C. Perez, Can't sleep, clown will eat me, Chaz1dave, Cherkash, Chienlit, Chowbok, Chr.K., Cotterstock, Cs-wolves, Cybergothiche, D.M.N., DH85868993, Daverso, Dpv, DragonHawk, E.G., Edward Waverley, Elk Salmon, Ericd, Foxandpotatoes, Gallant g, Ganeshtv, Gzuckier, Hajor, Hede2000, Herranderssvensson en, Ian Dalziel, Immunize, Jbattersby, Jfitts, JillandJack, Jnc, John Anderson, Juan25, Jwissick, KaptenKrater, KnightRider, Knightserbia, KyleRGiggs, Lapuwali, Lomedae, Lustigson, Madridista92, Majin Izlude, Marcus22, Mark83, Maroniofgrandprix, Matthead, Mjackson, Morio, NaBUru38, NightCrawler, Officially Mr X, Ortcutt, Paddyslacker, Pc13, Pelago, Peroxwhy2gen, Pinizicc65, Poo5, Pyrope, Rdsmith4, Readro, Recury, Red marquis, Redrose64, RekishiEJ, Rich Farmbrough, Rjwilmsi, Rosp, SM, Shadoks, SharShar, SidP, Smalljim, Sporti, Staffno, Ted Wilkes, The Tramp, Thincat, Tide rolls, Timc, Toraiannguru, Typ932, Volker89, Woohookitty, Yurek88, Zvonko, 75 anonymous edits

World_War_I *Source*: http://en.wikipedia.org/w/index.php?title=World_War_I *Contributors*: $1LENCE D00600D, *drew, -- April, -Inanna-, -Majestic-, -sd-, .marc., 007cat, 09dartnellalexander, 0kdal, 0zymandias, 119, 1234r00t, 137.205.8.xxx, 13dble, 172, 195.13.162.xxx, 195.149.37.xxx, 1970gemini, 1exec1, 209.20.135.xxx, 22344, 2601andrew, 27 chicks, 2812, 2d dogg, 3Jane, 3mta3, 62.253.64.xxx, 75th Trombone, 8digits, 96T, 98smithg2, A Nobody, A Quest For Knowledge, A Werewolf, A. S. Aulakh, A.Cython, A.Garnet, A.S. Brown, A.h. king, A1m1s1g1, AHM, AKucia, ALargeElk, AR, AThing, Aar, Aaron DT, Aaron Schulz, Aaron brown, AaronCBurke, Aaronpegram, Aaronsclee, Aaronstj, Abby 92, Abe Lincoln, Abeg92, Abrookins2000, Academic Challenger, Accurizer, Acebrock, Acefox, Acegikmo1, Achmelvic, Achnash, Acyso, Adam Cuerden, AdamRaizen, AdamRetchless, AdamSmithee, AdamantlyMike, Adambondy, Adambro, Adashiel, Adhib, Adrian, Aeon1006, Agamemnon2, AgnosticPreachersKid, Aguasde13013, Ahmad.ibn.as.Sayyid, Ahoerstemeier, Aim Here, Aimsme, Airborne84, Airbornelawyer, Airconswitch, Ajaxkroon, Ajraddatz, Ajw18, Akamad, Akhil 0950, Aki009, AkifSarwar, AksamitSK, Aksi great, Al.Glitch, Albanman, Albrecht, Aldahiri1, Aldis90, Ale jrb, Alensha, Alephh, Alex Kinloch, Alex1985, Alex404, Alex:D, AlexPlank, Alexlange, Alexmilt, Alexultima, Algabal, Algebraist, Ali K, Alison, Allmightyduck, Alltat, Alphachimp, Alrasheedan, Altau, AltayAtli, Altenmann, Altes, Altuga, Alunar, Alvin Lee, Alvinward, Amacher, Amazing Flash, Amazon10x, AmiDaniel, Amitprabhakar, Ammar shaker, Amrush, Anam Gumnam, Anand Karia, Anastrophe, Ancheta Wis, Andre Engels, Andreasegde, Andres, AndrewHowse, Andrewdev, Andrewlp1991, Andrewpmk, Andrewrost3241981, Andropod, Andy Marchbanks, Andy120290, AndyZ, Andycjp, Andygx, Andyjsmith, Angela, Angelo Somaschini, Anger22, AniBunny, Animaly2k2, Animum, Ankhurdhar, Ankit jn, Anne97432, Anomie Schmidt, Anon user, Anonymous editor, Anotherclown, Antandrus, Anthony Winward, Antidiskriminator, Antodav2007, Aodhdubh, Appleboy, Aprilbd, Aqs34, Aquatics, Arabani, Aranel, Aranherunar, Aratuk, Arcadie, Arcadina, Arch dude, Archanamiya, Ardric47, Arkalochori, Arni90, Arno, ArphaxadHunter, Arsonal, Art LaPella, Aruffo, Aruse, Arwel Parry, Arx Fortis, Asarelah, Asarlaí, Asiaticus, Asnac, Aspinelli, Assed206, Assyria 90, AstroNomer, Astroview120mm, Ataleh, Athenean, AtilimGunesBaydin, Atlan, Atletiker, Ato, Atropos, Attilios, Aude, Audrey, Auhlman, Auntof6, AusJeb, AutoGeek, Avant Guard, Avaya1, Avenue, AvicAWB, Avidius, Avraham, Avs dps, AxelBoldt, Axl, Axxxion, AzaToth, Aznboi168, Aznerd11, B, BBnet3000, BGTopDon, BOARshevik, BW, Bagel7, Bahamut Star, Baks, Balcer, Bambuway, Bamstick, Bamzguitargurl, Banes, Bantosh, Baristarim, Barking1, Barriodude, Bart Versieck, Bart133, Bastin, Batch1928 44, Baxter9, Bbaerst, Bbatsell, Bbboy657, Bberoth, Bcatt, Bci2, Bcorr, Bearcat, Bearishpeter, Beautyandbrains710, Bebek101, Bedivere.cs, Before My Ken, BeholdMan, Bejinhan, Beland, Belginusanl, Bellhalla, Belligero, Ben D., Ben davison, Ben-Zin, Bender235, Benjaminrobertfield, Benna, Benw, Benzo, Berean Hunter, Berek, Betacommand, Bethaso, Bevo, Beyond My Ken, Beyond silence, Bfx0, Bg007, Bhadani, Bhawani Gautam, Bhumiya, Bibliomaniac15, Big Bird, Big Iron, BigFatBuddha, Bigar, BiggKwell, Bigger digger, BigglesTh9, Biles1984, BillShurts, Biruitorul, BitterMan, Bizso, Bizub4, Bjw3, BlackDart D, Blackangel25, Blackhawk003, Blackknight12, Blackrx, Blackvault, BlaiseMuhaddib, Blaxthos, Blee145, Bleh999, BlisteringFreakachu, Blondito, Blue520, BlueGoose, Bluedesk, Bluemoose, Blupping, Bluseychris, Bmf 51, Bob bobato, BobWeaver112, Bobblehead, Bobblewik, Bobcats 23, Bobo192, Bobsmith76, Bodhran, Bodnotbod, Bogdan, Bogey97, BokicaK, Bolgeface, Bones13X, Bongwarrior, Bookandcoffee, Borg*Continuum, Boris Godunov, Bornintheguz, BostonMA, Boundarylayer, Bourquie, Boyohio02, Bozgo, Br43402, Brackenwood, Brado32003, BraikoT, BrainyBabe, Brandmeister (old), Bratsche, Bravado01, Brendan Moody, Brendenhull, Breno, Brenont, Brewcrewer, Brian the Fat, Brian0918, Brianhe, Brianpie, Brianski, Bridgeplayer, Brighterorange, Brisvegas, Britishliregts, Brnzwngs, Broberds, Brocky44, BrokenSegue, Bronks, Brookelittle, Brovnik, Brrrrrett, Bruce1ee, Bruce89, Brutannica, Bryan Derksen, Bryson109, Bsimmons666, Bt8257, Btphelps, Bu2m5dgw, Buchanan-Hermit, Bucketsofg, Buckshot06, Buldri, Bull Market, Bummerboy, BurlyPoet, Burntsauce, Burto88, Burwellian, Butros, Button2, Butwhatdoiknow, Bwidi, C12ypt, CART fan, CJK, CTZMSC3, Cachedio, Cactus.man, Caelarch, Caesarsgrunt, Caesura, Cafzal, Caglarkoca, CalJW, CalMc90, Calanon, Callainen, Calmypal, Caltas, Camarcus, Cambalachero, Camo99, Camp3rstrik3r, Camryl, Can't sleep, clown will eat me, Canadia, Canadian-Bacon, CanadianCaesar, Canderson7, CanisRufus, Canuckian89, Caper13, Capt Jim, Captain deathbeard, CaptainFugu, CaptainP, Carabinieri, Carl Logan, Carlaude, Carlwev, Carnildo, Carom, Carrick roads, Casmith 789, Casper2k3, Cassivs, Catgut, Causa sui, Cautious, Cbbv34, Cbdorsett, Cbrown1023, Cdamama, Cedders, Cedricbear, Cedrus-Libani, Celebration1981, CeleritasSoni, Celestra, CellyPhono, Centomo86, Central Data Bank, Centralpanthers, Centrx, Cerceole, Cerealkiller13, Cglassey, Chachob, Chainclaw, Chal7ds, Chaldean, Chameleon, Chaosdruid, Charles Matthews, CharlotteWebb, Charminultra123, Chase me ladies, I'm the Cavalry, Chavorz, Chawol, Chazz88, Chederman69, Cherkash, Chessy999, Chiguel, Chile, Ching, Chkiss, ChoraPete, Chrfis, Chris 42, Chris Roy, Chris the speller, Chris-marsh-usa, Chris1219, Chris7121992, Chrisdab, Chrism, Christian Historybuff, Christian List, Christophenstein, Christopher Parham, Christopher denman, Chriswiki, Chubleathem, Chump Manbea, Chupon, Cico51, Citra28, Civil Engineer III, Cjnm, Ck lostsword, Ckatz, Ckoicedelire, Clarityfiend, Clay70, Cleatus93, Clemwang, Cliché Online, Climie.ca, Clintville, Cloak Reaver, Cloud10, Cls14, Coachwaves, Cobey13, CodeCarpenter, Codik, Coemgenus, ColinClark, Coljac, Colonel111, Coloneloftruth, Commodore2468, CommonsDelinker, Commuood, Computerjoe, Confuciou, Connerite false mythology, Connorjohndyer, Conversion script, Coolboy223, Coolguy1351, Copperchair, Corpx, Corto lu, Corvatis, Corvus coronoides, Costesseyboy, Costoa, Cotepoo, Coughinink, CrackWilding, Crazycomputers, Creature, Credema, Crownjewel82, Crudcrudcrud, Crunkdaddy, CryptoDerk, Crzrussian, Cube2000, CuffX, Curps, Cutler, Cvieg, Cwill151, Cwkmail, Cyberevil, Cyktsui, CylonCAG, Cyrenaic, D1111, DARTH SIDIOUS 2, DECKitBRUISEit, DHN, DITWIN GRIM, DJ Clayworth, DUMJOO, DVD R W, DW, DWRYATE, Dacbook, DaddyWarlock, Daddywalrus, Daffy duck, DagosNavy, Dale Arnett, Dalisback1, Dan Barringer, Dan100, DanKeshet, DanMS, Dana boomer, Danbloch, DancingPhilosopher, DandyDan2007, Daniel Brockman, Daniel Garcia-Galili, Daniel Olsen, DanielCD, Danski14, Dark-hooded smoker, DarkFalls, DarkFarmer, DarkHorizon, Darkscholar789, Dart Kietanmartaru, DarthRad, DarthShrine, Davadr, Dave Harris, Davedx, Davee, Davenbelle, Davewho2, Davewild, David E Welsh, David Gerard, David Kernow, David Parker, David R. Ingham, David Underdown, David.Mestel, David.Monniaux, David909, DavidA, Davidcannon, Davivalle, Db099221, Dbrodbeck, Dcandeto, Dchall1, Ddspell, Dduck, DeLarge, Deadmaster, DearPrudence, Death2All, Deathshadow391, Deb, Deborahjay, Debresser, Dedonite, Deeb, Deenoe, Dejudicibus, Deleted1231, Delighted eyes, Delirium, Delldot, Deltabeignet, Delusion23, Demmy, Den fjättrade ankan, Dendodge, Denihilonihil, Dennisaurus, Deorum, DerHexer, Derek Ross, Detroit4, Devin M, Dewritech, Dfdc, Dfhdfhukdbvgfg, Dfrg.msc, Dhartung, Dhruvsha, Diatarn iv, Diderot, Dietary Fiber, Digitalme, Dino, Dinoguy2, Dinohunter999, Dinopup, DionysosProteus, DirkvdM, Disavian, Discospinster, Dissident, Distortedlojik, Djidash, Djmutex, Djsports0831, Dlbarber, Dmn, Dna-webmaster, Doc Quintana, Doc glasgow, DocWatson42, Doltna, Domdowns, Domino theory, Donald j axel, Donswaim, Doradus, Dormskirk, Dorvaq, DoubleDeuce721, Doughboy, Dpaajones, Dpm, Dpmwuzzlefuzzle, Dputig07, Dr. Thug, Dracion, DragonHawk, Dragonire, Dreadstar, Dreamafter, Drestros power, Drknkn, Drmies, Drogo Underburrow, Drummerboy101, Drunkenmonkey, DuKot, Duck of Luke, Dude1818, DuncanHill, Durin, DurotarLord, Durova, Dv82matt, Dwp13, Dycedarg, Dynex811, Dysepsion,

DéRahier, ESkog, ESommers, EWS23, Eaefremov, Ealster2004, EatAlbertaBeef, Ebyabe, Eclecticology, Edcreely, Edemaine, Edgar181, Editor at Large, Edkollin, Edsanz, Edthed, Eduardo Sellan III, Edward, Edwy, Ehinson56, Ehistory, Ehwills, Eichlmat, Eizo87, Ekyygork, El C, El Cid, El gato verde, ElTyrant, Eleassar777, Elelmnelo, Elendil's Heir, Elf-friend, Elfguy, Eliz81, Ella Plantagenet, Elustran, Elvarg, Emberstone878, Emc2, Emccain, Enceladus, EngineerScotty, Enter The Crypt, EoGuy, Epolk, Eran of Arcadia, Eregli bob, Erendwyn, Eric learner06, Ericd, ErikHaugen, Ertz, Esaborio, Escape Orbit, EscapingLife, Esrever, Etm157, Eu-151, Eubulides, Eugene van der Pijll, Eupator, Eurosong, Euryalus, Evercat, Everyking, Evgenikovalev, Evil Monkey, Experimental Hobo Infiltration Droid, ExtraordinaryMan, Exuperantius, Ezra Wax, Ezrakilty, Eóin, F Notebook, FBM, FC Toronto, FClef, FETuriousness, FF2010, FISHERAD, FRANKENSTONED, Fabykot, Fact Checker, Factuarius, Fadix, Fael, Fah007, Fair Deal, Falconhurst1967, Falconus, Fang Aili, Farawayman, Fauck, Favonian, Fawcett5, Fdp, FeanorStar7, Feitclub, Felixpetrar, Fenevad, Fernsehturmaufzug, Ferrarifreak890, Fetchcomms, Feydey, Filip1507, Findingemo, Fish and karate, Flaremon, Flarn2006, Flauto Dolce, FlieGerFaUstMe262, Floaterfluss, Flockmeal, FloopyFlarg, Fluteflute, Flux.books, Flyguy649, Fmalan, FoamParty, Fol2choco, Folkward Jansma, Foobaz, For one soul, ForgottenManC, Formeruser-82, Forteblast, Fourthords, Foxinsox, FrYGuY, Fram, Francs2000, FranksValli, Frankwomble, Franky210, Frazzydee, Freakofnurture, Frecklefoot, Fred Bauder, FredLGibsonJr, Fredrik, Freebiegrabber, Freemarket, Frenkmelk, FreplySpang, Friedfish, Frietjes, Frogg, Froggyjumpoverlog, Frogman574, Froren, Frostyservant, FrozenPurpleCube, Fuhghettaboutit, Func, Funeral, Funnybunny, Funnyhat, Furiousdick, Furrysaint, Future Perfect at Sunrise, Fuzheado, Fuzzie, FvdP, Fvw, Fwed66, Fyyer, G-W, GABaker, GCarty, GDonato, GHe, Gabbe, Gaff, Gage123, Gaius Cornelius, Galactor213, GambitMG, Gamesax7341, Gamewizard546, Gang65, Gap, Garik, Garrettt102, Gary King, Garzo, Gazpacho, Gdo01, Gdr, Gebbie, GeeJo, Gene Nygaard, GeneralPatton, GeoffreyVS, GeorgeBuchanan, GeorgeLouis, Geossj5, Gerbrant, Gerolsteiner, Gerry Lynch, Get It, Ggugvunt, Ghepeu, Ghilz, Ghirlandajo, Ghost109, GhostofSuperslum, Giancarlos, Gidonb, Gikü, Gilliam, Gillyweed, Gimboid13, Ginkgo100, Giraffedata, Gito, Gjenvick, Gkoehler70, Glacialfury, Glaug-Eldare, Glen, Gligan, Gman124, Gmaxwell, Gmcole, God Bless America, God of guitar, Gogo Dodo, GoldRingChip, GoldenMew, Goodoldpolonius2, Goodtimber, Gosgood, Gossmaaf, Gpeilon, Gr8opinionater, Gracenotes, Graculus, GraemeL, Grafikm fr, Graham Chapman, Grammarmonger, Grandgrawper, Grandmastergrawp, Granpuff, Grant65, Grasshoper, Grawp, Green Cardamom, Green Squares, Green01, GreenChild, GregRustFan, Greyengine5, Groogle, Ground Zero, Groundsquirrel13, Grover9, Grstain, Grumpyapp, Gsl, Gtg204y, Guaka, Guanaco, Guigui169, Gun Powder Ma, Gunmetal Angel, Gunwarband, Guoguo12, Guppie, Gurch, Gurubrahma, Gurvon, GusF, Gusreed, Guy Hatton, Guyjohnston, Guywh0sitsbehindphilip12, Gwernol, Gymnophoria, Gzkn, H. Jonat, HARRY POTTER, HJV, Haber, Hadal, Hadigonzalez25, Hadžija, Hairhorn, Hairy Dude, Haldraper, HalfShadow, Halibutt, Hall Monitor, Halogenated, HandThatFeeds, Handy wall link, HannahBana, Hansm77, HappyCamper, HappyDog, Harikishore, Harizotoh9, Harjk, Harkathmaker, HarrisonB, HarryHenryGebel, Harryboyles, Harsimaja, Harthacnut, Hashshashin, Hatch68, Hawaiian717, Hayabusa future, Hbmarlene, Hbuchanan, Hchc2009, Hcheney, Hdt83, Headbomb, Headphonos, Healthinspector, Heathfieldcc, Heavymanplease, HeikoEvermann, HeinzzzderMannn, Hellbound Hound, Hellinon, Hello32020, Henrik, Henry Flower, HenryLarsen, Hephaestos, Hephæstion, Herb45, HereToHelp, Heron, Herostratus, Herr Anonymus, Hesperian, Hestemand, Heyitsalexander, Heyyou17, Hibernian, Hide&Reason, Hideousdarren, High King of the Noldor, Historian932, History13, Hit45, Hmains, Hmoleman00, Hohns3, Homagetocatalonia, Howcheng, Hpmachine, Hrjohnson10, Hu12, HubertCumberdale, Hughstew, Hugo999, Hurricane Angel, Hurricane111, Husond, HussainAbbas, Hut 8.5, Hux, Hydrargyrum, Hydrox, I already forgot, I.M.S., IBen, IIIraute, IJA, IRT.BMT.IND, ISpinksy, IZAK, Iaaasi, Iain marcuson, Iakane49, IanW, Ians, Icarus, Icsunonove, Idaltu, IdreamofJeanie, Igilli, Ignatzmice, Ihuxley, Ikiroid, IkonicDeath, Il palazzo, Illegitimate Barrister, Iluvcapra, Ilya1166, Imaglang, Imaninjapirate, Imperdimper, Impossiblepolis, Imyourfoot, Indoles, Indon, InfernalPanda, InfinityAndBeyond, Infrogmation, Inkypaws, Intelligentsium, Inter, IntrigueBlue, InvisibleK, Iota, Ipankonin, Irenectc, Iridescent, IrishJew, Irishguy, Iritakamas, IronGargoyle, Ironik kid, Irpen, Irregulargalaxies, Irrypride, Irvine, Is Mise, Iscmus, It Is Me Here, Itai, Italus, Itismee, Itsalive4, Itsmine, Ivanljig, Ivexxl, IvoShandor, Ivtectuner, IxK85, Ixfd64, Izehar, Izic739, Izno, J M Rice, J. Naven, J.R. Hercules, J04n, JCSantos, JFHJr, JForget, JFreeman, JHawk88, JIP, JK the unwise, JL 09, JLD, JMK, JONJONAUG, JPEriksson, JSpung, JTBX, JTN, JW1805, JY.public, Ja 62, Jaaches, Jab692, Jacek Kendysz, Jack Bethune, Jackanapes, Jackietang33, Jacoplane, Jaedza, Jagged 85, Jakeferrett, Jakswa, Jam1993, James Seneca, James086, JamesAM, Jamesadam739, Jamesmorrison, Jamesofur, JamieKeene, Jankyalias, Jao, Jason M, JasonAQuest, JasonB007, JasonM, Jasoncward, Jasonglchu, Jatder, JavaTenor, Jawms, Jaxl, JayEsJay, Jayjg, Jaymano, Jayron32, Jcmurphy, Jcullen751, Jcw69, Jdevine, Jdigangi, Jean-Jacques Georges, Jean.Miller, Jeff G., Jeff3000, JeffK1971, Jeffkw, Jeffrey O. Gustafson, Jegmwynn, Jellybean308, Jengod, Jennica, Jensbn, JeremyA, Jeremystalked, Jergen, Jeronimo, Jerzy, JesseHogan, Jfdwolff, Jh51681, Jhamez84, Jhami3, Jhawk1024, Jiang, JidGom, Jiddisch, Jigen III, Jim Sweeney, Jimblack, JinJian, Jiujitsuguy, Jj137, Jjeffrey, Jkelly, Jll, Jlwiki, Jm51, Jmasalle, Jmeppley, Jmlk17, Jmw0000, Jni, Jnk, JoJaysius, JoanneB, Joaopais, Joconnor, Joe Canuck, Joe N, JoeBlogsDord, JoeSmack, Joefromrandb, Joelr31, Joeseth1992, Joesnicecars, John, John E. FitzGerald, John K, John Millikin, John Paul Parks, John Quincy Adding Machine, John Riemann Soong, John Smith's, John Vandenberg, John of Lancaster, JohnCD, JohnI, JohnMac777, JohnWittle, Johnbibby, Johncmullen1960, Johngaffney, Johnleemk, Johnsemlak, Johnstevens5, Jojhutton, Jokerintrousers, Jombo, JonMiller, Jonathan.s.kt, Jonesy, Jooler, Joolz, Jordan117, Jordanperryuk, Jordanxxi, Joseph Solis in Australia, Josh Parris, Joshua Scott, Josias Bunsen, Jossi, Joy, Joyous!, Jpeob, Jpgordon, Jredmond, Jrt989, Jtdunlop, Jtopgun, Jtsoldier, Juangdsi, July4th, Jusdafax, Jusjih, JustFacts, Justin Mitman I, Jwkane, Jzcool, KConWiki, KGV, KPH2293, Ka Faraq Gatri, Kafziel, Kaiser Taylor, Kaisershatner, Kakofonous, Kampfgruppe, Kansas Bear, Karabekir, Karesu12340, Karl Stas, KarlFrei, Karlbrookes, Kate, Kate Phaye, Katefan0, Kayac1971, Kayuki16, KazekageTR, Kbeideman, Kbthompson, Kcordina, Keepdone, Keith Edkins, Keith-264, Kelisi, Kelsievans, Kelvinc, Ken Arromdee, Ken Gallager, Ken l lee, Kenman2, Kenmccue, Kenyon, Keopeli, Kerotan, Kevin, Kevin Murray, Kevin Yie, KevinMM, Kevinalewis, Kevlar67, Kewp, Kfc1864, Khalid Mahmood, Khaosworks, Khazar, Khemmingsen, Khoikhoi, Khukri, Kidace, Kieshabelvett, Kilrogg, Kimdime, Kimyu12, King of Hearts, Kingturtle, Kirill Lokshin, Kirkesque, KissL, Kissall, Kizildere, Kizor, Klaser, Klemen Kocjancic, Knight45, Knightserbia, KnowledgeOfSelf, Knows stuff, Knutux, Koffieyahoo, Konstable, Koro Neil, Kosebamse, Kotjze, Kowh, Kozuch, KraMuc, KrakatoaKatie, Krankman, Krellis, Krich, Kris159, KrisDIM, Krn10919, Kross, Ksenon, Kshieh, Ktan91, Kuguar03, Kukini, Kungfuadam, Kungming2, Kuroi, Kurt Leyman, Kuru, Kurykh, Kutukagan, Kvladiko, Kwertii, Kwidner, Kyle sb, Kylebrennan1, L Kensington, L0b0t, LCpl, LOL, LOTRrules, LUH 3417, LWF, Labattblueboy, Laca, Lachlan Smith, Lacrimosus, Laddiebuck, Laisinteresting, Lakers, Lancevortex, Landr, Languagehat, Lanoitarus, Lapsed Pacifist, Lars T., Laurinavicius, Lawilkin, Lawrencekhoo, Lawrencema, LeadSongDog, Leandrod, Lecen, Lectonar, Lee1026, Leithp, Lemmey, Lemonade100, Lemonflash, Lensi, Leon rules, Leonard Vertighel, Leonariso, Leoni2, Leopold Stotch, Leszek Jańczuk, Leuko, Levineps, LiamE, Library time, Lifung, LightSpectra, Lightdarkness, Lightmouse, Ligulem, Lilac Soul, Ling8zhi, Linty man, Lisiate, LittleOldMe, Livajo, Llort, Llydawr, Logan, Logical2u, Logicalanswer, Longpatrol42, Lord Iffy Boatrace, Lord Voldemort, Lord khadgar05, Lorenzarius, Lothar von Richthofen, Lotje, Louis Do Nothing, Louis88, Lowellian, Lradrama, Lt.Specht, LtWinters, Ltwin, Lucano, Lucifero4, LukeS, Lukeisham, Lumingz, Luna Santin, Lupo, Luuva, Lx 121, M A Mason, M-le-mot-dit, M.O.X, M.nelson, MBK004, MC MasterChef, MECU, MER-C, MGRILLO, MJBoa, MONGO, MOTORAL1987, MPowerDrive, MRSC, MStar, MTG1989, MWAK, MZMcBride, MackSalmon, Mackay 86, Macschanger, Mactographer, Macukali, Madmagic, Maeglin Lómion, Maelnuneb, MaerlynsRainbow, Magbatz, Magister Mathematicae, Mai-Sachme, Mailer diablo, MajklDzekson, Majorly, MakeChooChooGoNow, Malcolm Farmer, Malleus Fatuorum, Malo, Man vyi, Mana Excalibur, ManagementF1, ManiF, Mapcat, Mapper76, MapsMan, Maralex334, Marasmusine, Marc Venot, Marc omorain, Marcpatt14, Mareino, Marek69, Marine 69-71, MarineCorps, MarkGallagher, MarkSutton, Markkawika, Markussep, MarshallBagramyan, Marskell, Martel ND, Master of Puppets, MasterHalco, MasterOfHisOwnDomain, Masterblooregard, Mat 61, Materialscientist, Mathew5000, Mathfan, Mathx314, Matilda, Matsuya, Matt Adore, Matt Crypto, Matt Yeager, Mattbell8808, Mattbuck, Matters123456, Matthew Husdon, Matthew Samuel Spurrell, MatthewVanitas, Matthewmayer, Mattisse, Mattmeskill, Mattratt9, Mattstan, Mattyh190, MaulYoda, Mausy5043, Mav, Max Density, Max-zbikowski, Maxf.m, Maximaximax, Maximilian Schönherr, Maximilli, Maximus Rex, Maxpowers12345, Mbossa, Mboverload, Mbr0052, Mbz1, Mc4932, Mcjanus, Mcoov, Mdeen, Meaningful Username, Meelar, Meeso, MegX, Megistias, Melchoir, Melsaran, MeltBanana, Meltonkt, Mephiles602, Mercury, Mersperto, Metaphysicus, Mewillundo, Mgiganteus1, Mhardcastle, Michael Devore, Michael Dorosh, Michael Essmeyer, Michael Zimmermann, Michaelminett, Michaelrccurtis, Michaelzeng7, Michig, Micro101, Mieciu K, Miesianiacal, Mightrane, MikailMoolla, Mikailus, Mike Rosoft, MikeLynch, Mikeheartstone, Mikemoral, Mikestone8, Mikey23, MilborneOne, MilesVorkosigan, Miller101, Mindspillage, Mindstar, Minesweeper, Minna Sora no Shita, Mintguy, Mirlen, Mischa83, Mishadeon, Misplaced canadian, Mista poe, Misza13, Mitlikespenies, Miyagawa, Mk5384, Mkpumphrey, Mlpearc, Mlwgsgis1487, Mm40, Mmortal03, Mo0, Moanzhu, Moby Dick, Mocirne, Modulatum, Moesian, Mokoko 24-7, Molly-in-md, Mollyclare, Molobo, Monkeyman, Mono, Monocrat, Moogin, Moonriddengirl, MooseHockey, Moreschi, Morevisit, Mormegil, Morrad, Morrisqc, Mortenhelles, Moshe Constantine Hassan Al-Silverburg, Moxy, Mozzerati, Mpatel, Mpfrenchy13, Mpking52, Mr Mulliner, Mr Stephen, Mr. Billion, Mr. Military, Mr. Penis Vandal, MrDolomite, MrFish, MrWeeble, Mrflu2, Mrmariokartguy, Mrmaroon25, Mrmiscellanious, Mrstonky, Mrtorrent, Ms2ger, Mschel, Msclguru, Msrasnw, Mswake, MuZemike, Mur525, Mushroom, Muzekal Mike, Mvaldemar, Mwanner, Mxn, Myanw, Mycroft7, Mynameisdomovoi, Mystichumwipe, Mzajac, Mzk, Mário, Möchtegern, N328KF, NFactorial, NIN1337, NSK Nikolaos S. Karastathis, Nach0king, Nakon, Nanuck, Naohiro19, Naraht, NarayanGa, Nareek, Nate083, Nathan Johnson, Nathanjp, Natl1, Navy Blue, NavyHighlander, NawlinWiki, NazismAintCool, NebY, Neddyseagoon, Nelro, Neodammerung, NerdyScienceDude, NeroN BG, Netfrack, Neutrality, NewEnglandYankee, Nfitz, Nfleming, Nford24, Nibuod, Nick, Nick Number, Nick UA, Nick-D, NickisCool, Nickj, Nickwolf, Niclah99, Nicolas M. Perrault, Nigel Ish, NigelR, Nil Einne, Ninjaramo, Nintendude, Nirvana2013, Nirvana77, Nishkid64, Nitrogendragon, Nivix, Nixer, NjtoTX, Nk, NlynchN, Nmpenguin, No Guru, Nobody of Consequence, Nobunaga24, Noclador, Nog64, Nonagonal Spider, Nooby god, NorbertArthur, NorkNork, Northamerica1000, NorthernThunder, Nosferatublue, Notoriousdoc, Nowax, Noworld, Nsaa, Nsandwich, Ntsimp, NuclearWarfare, Numbo3, Nurg, Nv8200p, Nwe, Nydas, Nyenyec, Nyttend, Nzpcmad, OMGTANGERINES, OOODDD, Obarskyr, Object.toString(), Obli, Oblivious, Obradovic Goran, Ocee, Octopus-Hands, Oed, Ogbn, Ognir, OhanaUnited, Ohconfucius, Ohnoitsjamie, Okapi, Old Moonraker, Oleg-ch, Olegwiki, Oliver202, Olivier, Oliviu, Olorin28, Omicronpersei8, Onceonthisisland, OneVeryBadMan, Onlinetexts, Opabinia regalis, Opelio, Opes, Opiner, Orangemarlin, Orangutan, OrbitOne, Oreo Priest, Orii, Orlando098, Orpheus, Ortolan88, Osbojos, Osmanja, Otberg, Othello, Otto ter Haar, OttomanReference, Ottre, OwenX, Oyo321, Oyvind, P3Pp3r, PANONIAN, PBP, PDH, PDTantisocial, PFHLai, PHXsuns3434, PJBEAR13, PJM, PMLawrence, POTUS270, Pacquiao, Paine Ellsworth, Paleo dragon, Palica, Panairjdde, Pangeanempire, Papy 89, Parkerh7893, Parkwells, Parsecboy, Partizanfighter1944, Pascal.Tesson, Pat Payne, PatMcClendon, Patar knight, Pathoschild, Patricknoddy, Patrolmanno9, Pats1237, Patstuart, Paul 012, Paul August, Paul Barlow, Paul Benjamin Austin, Paul.h, Paulcicero, PaxEquilibrium, Peachsncream, Pearle, Pedant17, Pedestrian06, Pedro Ximinez01, Peemil, Pembers, Pennybanks2, Pepo13, Perceval, Perdy80, Peregrine981, Peruvianllama, Peter Atwood, Peter Bell, Peter Grey, Peter R Hastings, Peter Tribe, Peter-File, PeterisP, Peterspage, Petri Krohn, Pevarnj, Pewwer42, Pgan002, Pgk, Phantom784, Phantomsteve, Phil alias Harry, Phil153, PhilKnight, Philip Baird Shearer, Philip Trueman, PhilipIsPDR, Philippe, Phoenix Hacker, Phoenix2, Phonemonkey, Photonikonman, Phydend, Physchim62, PiCo, Pietdesomere, Piggman248, Pigman, Piledhigheranddeeper, Pilotguy, Pinethicket, Pinnecco, Pinzo, Piotr Mikołajski, Piotrus, Pirkid, Pizza Puzzle, Pjamescowie, Pjg12, Pklinker, Planders, Plantago, Plasma Twa 2, Plasmic Physics, PlatinumX, Plau, Plerdsus, Pluma, Pmcelroy77, PochWiki, Poetaris, Pogoman, Pokedork7876, Pol098, Politepunk, Politics rule, Polotet, Polycarp, Polylepsis, Polylerus, Prashanthns, Predius, Preetikapoor0, Prelle, PrestonH, PrimeCupEevee, PrimeHunter, Private Butcher, Prodego, Professor Ninja, Proger, Protophobic, Psantry, Pseudo-Richard, Pslide, PsyberS, Psychonaut, Publicus, Pudupudu, Pulsemeat, Purpleturple, Pustelnik, Putush, PuzzletChung, Pyrotheblob, Q33q3, Qampunen, Qasibr, Qertis, Qman72, Quadell, Quebec99, Quibik, Quintessent, Qutezuce, R-41, R. S. Shaw, R9tgokunks, RJHall, RJaguar3, RPHKUSA, RSStockdale, RaCha'ar, Raccoonsrevenge, Radiotelemetry, Ragout, Raguleader, Rande M Sefowt, RandomP, Randomdude888, Randompiggy1, Randomtime, Rangek, Rangerwave, Rangi42, Raphater, RashersTierney, Rasmusdf, Raul654, Raven4x4x, Ravik, Raymond Palmer, Raywasp, Rbrwr, Rdsmith4, Reach Out to the Truth, Rebel Redcoat, Recato, Reconsider the static, Red Director, Red1001802, RedSpruce, RedWolf, Reddi, Redhead911, Redsoxrichardj, Redvers, Regara MkII, Reidmeister, Reisender, Remuseum, Renamed user 145, Renata3, Retired user 0001, Retired username, Rettetast, Reuvenk, Revolución, RexNL, Rexroad, ReyBrujo, Rfl, Rheo1905, Rhobite, Rhombus, Riana, Rich Farmbrough, Richard Arthur Norton (1958-), Richard Ye, Richard temple, RichardF, Richardaedwards, Richfife, Richieice, RickK, Rickyrab, Riddley, Rikuharts, Rillian, Riotforlife, Ripberger, Risker, Rjensen, Rjwilmsi, Rklawton, Rl, Rmhermen, Rmosler2100, Robert K S, RobertG, Robertgreer, Robertvan1, Robhd, Robin63, Robinmcghee, Robtl400, Rocastelo, Rocket71048576, Rocxror, Rodrigo Cornejo, Rofl, Roger Davies, RohanDhruva, Rokarudi, Romanm, Ron1423, Ronank, Ronbo76, Ronnotel, Rorschach, Rory096, RoslynSKP, RossPatterson, Rossrs, Roy Al Blue, RoyBoy, Royalguard11,

Royboycrashfan, Rpab, Rpeh, Rr parker, Rrburke, Rrius, Rsage, Rschroev, Rsyungul, RttlesnkeWhiskey, Ruby2010, Ruddy9hell, RudyB, Rueben lys, Ruggenberg, Ruhrjung, Rune.welsh, Rupert Horn, Russ3Z, RussBlau, Russia Moore, Ruzgar, Rwboa22, Rwestera, RxS, RyanGerbil10, Ryanhupka, Ryoung122, Ryryz666, Ryulong, SFX 1, SG, SGGH, SJP, SNIyer12, SSman07, ST47, SU Linguist, Sadads, Sade, Sailingstefan, SaltyPig, Sam Blacketer, Sam Golden, Sam Korn, SamB135, Sambian kitten, Samsara, Samuel Salzman, Samuel Webster, Samuella, Sandgem Addict, Sangil, Sango123, Sannse, Santolinek, Sargeras, Saros136, Sasquatch, Saturn070, Savant13, Savidan, Say1988, Sayden, Sbandrews, Scartboy, Schellack, Schmerguls, Schnell, Schroeder74, Schzmo, Science4sail, Scientz, Scifiintel, Scipius, Sciurinæ, Scolaire, Scorchy, Scoresby, Screwupwikipedia, Scruffy4903, Sdicht, Seabhcan, Sean Ray, Sean William, SeanWillard, Seanisthelizardking, Seanm687, Search4Lancer, Seb az86556, Sebesta, Secret (renamed), Securiger, Seki rs, Semperf, Semprini, SenorKristobbal, Seqsea, Seshi, Setanta747, Sevie586, Sextbeast, Sf46, Sfizzle, Sfly510, Sgdownard, Sgt. bender, Sgtconfig, Sgtcore, Shadooml, Shadow007, Shadow1, Shadowx5206, Shaidar, Shaidar cuebiyar, ShakingSpirit, Shaliron, Shalom Yechiel, Shanes, Shangrilaista, SharkD, Shawis, Sheldrake, Shem1805, Shenme, Shiaulteyr, Shimgray, Shinkz, Shipnerd62962, Shirimasen, Shoemaker's Holiday, Shoessss, Shpangler, Shreshth91, Shsilver, Shushruth, Siafu, Sietse Snel, Sifaka, Signaleer, Signalhead, Signsolid, SimonATL, SimonP, Simple Bob, Simply not edible, SinlordClaudia, Sir Intellegence, SirIsaacBrock, Sith92, Sjakkalle, Ske2, SkerHawx, Sketter101, Skier Dude, Skinny87, Skirmisher1331, SkonesMickLoud, Skookum1, Skyring, Skysmith, Sl1982, Slark, Slasherezz, Slatersteven, Sleigh, Sleptrip, Sliggy, SlimVirgin, Slovenski Volk, Sls9786, Slyguy, Sm8900, Smileyface11945, Smoketwojoints, Smokizzy, Smyth, Snapperman2, SnappingTurtle, SnowFire, Snowmanradio, Snowolfd4, Snoyes, SoLando, Soldaatvanoranje, Soler97, Solipsist, Solopiel, Some guy, SomeStranger, Someguy1221, Someguy321, Someone else, Someoneinmyheadbutit'snotme, Someuser69, SonOfJackass, SonicAD, Sonicdrewdriver, Spacevezon, Spamfactor, Spangineer, Spanglej, Sparky1895, SpeDIt, Specs112, Speight, Spellcast, Spellcheck, Spellcheck8, SpencerCS, Spettro9, Spinnaker gybe, Splash, Spliffy, Splinter1044, SpookyMulder, SpuriousQ, Spursnik, Squiddy, Srajan01, Srich32977, Srikeit, Srinivas, Srnec, Srushe, Ssimsekler, Ssmith619, Sstennett, StAnselm, Staberinde, Stan Shebs, StanZegel, Stargoat, Stears159, Stears81, Steel, Stefkornaros, Steinsky, Stephen G. Brown, Stephen Gilbert, Stephenb, Stereotek, Steve, Steve03Mills, Steve2011, Stevelomaka, StevenWT, Stevenscollege, Stevertigo, Stevooo, Stilgar135, Sting, Stmccoy, StoneProphet, Stor stark7, Str1977, Stratman, Straw Dogberry, Struthious Bandersnatch, StuRat, Stw, Suck12, Sumergocognito, Superhelix, Superlogan20, Superslum, Superswade, Superwolf89, SurrenderUK, Sus scrofa, Susurrus, Svensson, Svmich, Svnty, Swamilive, Swarm, Swedebu, Swedish fusilier, Swid, SwnyMac, Sybertronic, Sylvain1972, SyntaxError55, Szopen, Szquirrel, T-dot, TGregory, TJ Spyke, TJDay, TJRC, TRAJAN 117, TSO1D, TShilo12, TUF-KAT, Tabib, Tabletop, Tacoman2, Taemyr, Tafantic, Tagishsimon, TalentedMrRipley, Talkmuchlater, Talon Artaine, Tanaats, Tangotango, Tannin, TarikAkin, Tariqabjotu, Tarkata14, Tarquin, Tarret, Tasc, Tascha96, Tawker, Tazmaniacs, Tbhotch, TempestCA, Tempodivalse, Tempshill, Tenma06, Tenmei, Terence, Terps, TerryBP149, Tevzet, Tex, Textangel, Th1rt3en, Thannad, That Guy88, Thatdog, The Anome, The Disco King, The Enlightened, The Epopt, The Highest Tide, The King Of Gondor, The Land, The Man in Question, The Myotis, The PIPE, The Thing That Should Not Be, The Utahraptor, The Wookieepedian, The Wordsmith, The ed17, The tooth, The undertow, The wub, The-one, TheCoachZ11, TheKMan, TheMcManusBro, TheNeutroniumAlchemist, TheObtuseAngleOfDoom, TheRingess, TheTrojanHought, TheaterMarine, Theaznlaw, Thebigcurve, Thebomb667, Thecool58, Thedarkestone, Theevilempire666, Thegoodson, Thegreenj, Themanwithoutapast, Themoridian, Themusicking, Theon, Theone00, Theonlyedge, Therequiembellishere, Thewellman, ThoHug, Thomashwang, Thorpe, Three-quarter-ten, ThreeofFour, Thriller3000, Thtuskey, Thumperward, Tide rolls, Tifego, TigerShark, Tim Cox, Tim Q. Wells, Tim Starling, Tim bates, Tim103093, TimSE, Timc, Time Travler, Timeshifter, Timrollpickering, Tiniti, Tinss, Tipimad, Tito4000, Titoxd, Tiwonk, Tjunier, Tkil, Tkynerd, Tlrmq, Tmazhindu, Tmopkisn, ToastyKen, Tobby72, Tocath, Tokle, Tom Yates, Tom harrison, TomCerul, TomTheHand, Tombseye, Tommaylor, Tonberry King, Tony Sidaway, Tony1, Tony360X, Tophtucker, Topory, Torlib, Torqueing, Torve, Tourbillon, Tovarischtony, Toya, Tpbradbury, Transkar, Traroth, Trasman, Travelbird, Travmitz, Trecht, Trekphiler, Tresckow, Trevor MacInnis, Treyt021, Trickster206, Tricky Victoria, Trieste, TrogdorPolitiks, Tromboner7600, Troop350, TrulyTory, Trumpet marietta 45750, Truth1914, Truthkeeper88, Tstarl0425, Tuckerresearch, Turnhout, Tutmosis, Twanebo, Twinsday, Two Bananas, TwoOneTwo, Tycho, Tyjay06, Tyler Durgen, Typer 525, USA02, UberCryxic, Udora, Ugen64, Ugur Basak, UltimaRatio, Ultramarine, Ultratomio, Umsteadj, Umumu, Unbound123, Uncle Scrooge, UnicornTapestry, Unreal7, UpdArch, Upi, Upnoggdibo, Uris, Uronmyhitlist, Urzeitlich, User27091, Ute in DC, Uuchie, V, V3r71calh0r1z0n, VVPushkin, Vagodin, Valentinian, Valisk, Valrith, Vamink, Vamsilly, Van helsing, VanSisean, Vanish3, Vanished 6551232, Vanished User 1004, Vanished user 03, Vanished895703, Varlaam, Vary, Vazor20X6, VegaDark, Vendettax, Ventur, Ventus24, Venu62, Vercillo, Verrai, Vervin, VetMax, Vgy7ujm, Victor falk, Viewfinder, Vinn0r, Violetriga, Viridae, Viriditas, Vitriden, Vizu, Vocoindubium, Volker89, Voyevoda, Vrenator, Vroman, Vzbs34, W beattie, WAvegetarian, WED Imagineer, WODUP, Wafulz, Waggers, Wahabijaz, Wallie, Wally, Wally Wiglet, Walter, Wapcaplet, Wardog, Warrior on Terrorism, Wavelength, Wayward, Wbutler1, Welsh, Werdna, Westenra, Wewelsburg, Wham Bam Rock II, Whatsgoodbro, Whisky drinker, Whiskymack, Whisman, WhiteWriter, Whiteroll, Whoda, Whoosher, Whouk, Wierd al 101, Wiikipedian, Wik, Wiki Raja, Wiki alf, Wiki libs, Wiki-Ed, WikiLeon, WikiTikix, Wikibofh, Wikibout, Wikieditor06, Wikimandia, Wikipedian1234, Wikipediatastic, Wikispam, Wildthing61476, Wilhelm Ritter, Will Beback, WillOakland, Willdasmiffking, William Pietri, Willy, your mate, Wilybadger, Wimt, Wingspeed, Winick88, WinterSpw, Witchzilla, Wizardman, Wmahan, WolfmanSF, Wombatcat, Woody, Woohookitty, Woolfy123, Wooster boy, WordsExpert, Wrp103, Wutsje, Wwheaton, Wwoods, Wyrdlight, XLerate, XavierGreen, Xaxafrad, Xennik, Xevious, Xezbeth, Xiahou, Xiaphias, Xiong Chiamiov, Xkorean bbqx, Xtreambar, Xyzzyva, Yaddar, Yaf, Yahel Guhan, Yakuzai, Yama, Yamamoto Ichiro, Yandman, Yanksox, Yansa, Yaris678, Yasis, Yath, Yattum, Yellowoboe, Yelyos, Yemal, Yev900, Yfjonas, Ymal31, Yngvadottir, YoYoXehcimalYoWikiWikiWikiWhat, Yousou, Youssefsan, Ypacaraí, Yuckfoo, Yugioh1126, Yummy123, Zachary, Zaf, Zazzer, Zealander, Zeb543, Zedla, Zerotonin, Zharmad, Zibart, Zigamorph, Ziggy3055, ZmiLa, Zmilot, Zocky, Zoicon5, Zoken4, Zondor, Zreeon, Zsero, Zsinj, Zucchinidreams, Zurishaddai, Zzuuzz, Ævar Arnfjörð Bjarmason, Σ, Александър, Борис Романов, Камень, Свифт, , තමිල් විජයකර, €p a , , weizhe, 4686 anonymous edits

Hydraulic_brake *Source*: http://en.wikipedia.org/w/index.php?title=Hydraulic_brake *Contributors*: 4twenty42o, Aaronw, Alansohn, Andrew Swallow, Andy M. Wang, Avalon, Biscuittin, Bob350chev, Bobo192, Bookgrrl, Capricorn42, ChrisHodgesUK, Crazynas, Dannyc77, Discospinster, Ejay, Enviroboy, ErrantX, Fieldday-sunday, GB fan, Gilliam, GraemeL, Graham87, Gurch, Homebuilding, Hooperbloob, J.delanoy, Jacob805, Jamoche, Jbarnby, John254, KDS4444, Kymacpherson, LeadSongDog, Llywelyn, LorenzoB, LukeE94, LukeWiller2, Markiewp, Mtnsource, Orange Suede Sofa, Peter Horn, Pinethicket, Pseudov00, Redvers, Rogerzilla, Shoefly, Soeill, Sonett72, Stephenb, Teutonic Tamer, The Thing That Should Not Be, The wub, Tjn1976, Tom harrison, Travelbird, VMS Mosaic, Versus22, Vipinhari, VolatileChemical, Wayne, Wizard191, Ytrottier, 124 anonymous edits

Tours *Source*: http://en.wikipedia.org/w/index.php?title=Tours *Contributors*: 0, A dullard, Ahn J, Aka042, Alarics, Aleph Kaph, Allmhurach, Andre Engels, Angelo.romano, Angusmclellan, Anlace, Antandrus, Aranherunar, Ariobarzan, Arka Voltchek, Arthena, Attilios, Axeman89, Backspace, Balivers, Bfx12a9, BigDukeSix, Bigbeast1, BlueSquadronRaven, Bob f it, Bob247, Bobo192, Bobrayner, BorgQueen, Brandnewuser, CIreland, Caltas, Camelia11, Can't sleep, clown will eat me, Capricorn42, Captain scarlet, CardinalDan, Chameleon, Charlie678, Chasm4963, Cntras, DW, Da Joe, Daderot, Dan100, Darth 100, Darwinek, Didie (usurped), Discospinster, Douglasfrankfort, Eamoniski, Edbrims, Edward Grefenstette, Enchanter, Encyclopedia77, Enki H., Epsilon60198, Eric-Wester, Europe22, Ev, Excirial, Felix Folio Secundus, FlyingToaster, Fozi999, Fram, Francoisletiec, Franz22, FreD, Frietjes, Frokor, Ftpeeco, Funnyfarmofdoom, Gifðas, Grafen, Guill37, Gurchzilla, Hadal, HeartofaDog, Hede2000, Himalayantravelguide, History2007, I I OIO I I, Ihatemylifesuicide, Ikiroid, Infrogmation, J.delanoy, JHK, Jaardon, Jauhienij, JillandJack, Jim10701, Jimpi, Joao Xavier, Johncmullen1960, Johnnywalterboy, Joseph Solis in Australia, Julius.kusuma, Katieh5584, Kenneth131, Kiwipete, Klausi, Ksnow, L1A1 FAL, Lars bohinen, Lastemperor8899, LeilaniLad, LessHeard vanU, Lexor, Lfh, Lightmouse, Lofty, Luigi-ish, Lupin, MPF, Magnus Manske, Marek69, Markussep, Materialscientist, MichaelTinkler, Mikeo1938, Mr. Wasey, N5iln, Namiba, NativeForeigner, Neutrality, Nika 243, Ninly, Nk, Notafish, Ohconfucius, Ohnoitsjamie, Olivier, OllieFury, OwenBlacker, Ozeye, PCHS-NJROTC, Pearle, Piano non troppo, Plasticspork, Playclever, Poccil, Polina Parr, Porcher, Pseudomonas, QazPlm, Quintote, R'n'B, Raidon Kane, Rama, Rich Farmbrough, Rje, Rochelimit, Rrburke, Russian Luxembourger, Rusty2005, Ryebonfire, Sam Hocevar, Schzmo, Seraphimblade, Severo, Silver urn, Silversmith, Skunkboy74, Slawojarek, Snowolf, Specter01010, Squids and Chips, Steinsky, Stephenb, Struway2, TBHecht, Tango7174, Tarquin, Tatmouss, Tenegri, Tergum violinae, TheRanger, Themfromspace, Thomas Arelatensis, Thunderboltz, Tigga en, Twooars, UltimaRatio, Underwaterbuffalo, Vanished user 03, Vdnf, Vegaswikian, Vince37, Vrenator, W4chris, Waa03, Wameya, Wetman, Wiki alf, WikiLaurent, Woohookitty, Wsphar, Zacatecnik, 259 anonymous edits

Image Sources, Licenses and Contributors

File:Flag of Italy (1861-1946).svg *Source*: http://en.wikipedia.org/w/index.php?title=File:Flag_of_Italy_(1861-1946).svg *License*: unknown *Contributors*: User:F l a n k e r

File:Flag of Monaco.svg *Source*: http://en.wikipedia.org/w/index.php?title=File:Flag_of_Monaco.svg *License*: unknown *Contributors*: User:SKopp

File:Flag of France.svg *Source*: http://en.wikipedia.org/w/index.php?title=File:Flag_of_France.svg *License*: unknown *Contributors*: Anomie

File:Flag of the United Kingdom.svg *Source*: http://en.wikipedia.org/w/index.php?title=File:Flag_of_the_United_Kingdom.svg *License*: unknown *Contributors*: Anomie, Mifter

File:Circuit de la Sarthe track map.svg *Source*: http://en.wikipedia.org/w/index.php?title=File:Circuit_de_la_Sarthe_track_map.svg *License*: unknown *Contributors*: User:Will Pittenger

Image:LeMans PitStop2.JPG *Source*: http://en.wikipedia.org/w/index.php?title=File:LeMans_PitStop2.JPG *License*: unknown *Contributors*: user:Eyone

Image:Aston Martin DBR9 24h200702.JPG *Source*: http://en.wikipedia.org/w/index.php?title=File:Aston_Martin_DBR9_24h200702.JPG *License*: unknown *Contributors*: User:Fabe56

Image:Le Mans 2008 Rolling start.jpg *Source*: http://en.wikipedia.org/w/index.php?title=File:Le_Mans_2008_Rolling_start.jpg *License*: unknown *Contributors*: Wonker Wonker from London, United Kingdom

Image:Lemans Circuit Bugatti.JPG *Source*: http://en.wikipedia.org/w/index.php?title=File:Lemans_Circuit_Bugatti.JPG *License*: unknown *Contributors*: user:Eyone

Image:Circuit de la Sarthe track map.svg *Source*: http://en.wikipedia.org/w/index.php?title=File:Circuit_de_la_Sarthe_track_map.svg *License*: unknown *Contributors*: User:Will Pittenger

Image:1923Lemans.JPG *Source*: http://en.wikipedia.org/w/index.php?title=File:1923Lemans.JPG *License*: unknown *Contributors*: Erika39, Korrigan, Loveless, Man vyi, Spyder Monkey, 1 anonymous edits

Image:Renault Alpine A443 Donington 2007.jpg *Source*: http://en.wikipedia.org/w/index.php?title=File:Renault_Alpine_A443_Donington_2007.jpg *License*: unknown *Contributors*: Darren

Image:Porsche 962 1988 Le Mans at Silverstone 2007.JPG *Source*: http://en.wikipedia.org/w/index.php?title=File:Porsche_962_1988_Le_Mans_at_Silverstone_2007.JPG *License*: unknown *Contributors*: User:Pyrope

Image:Audi R10 1.JPG *Source*: http://en.wikipedia.org/w/index.php?title=File:Audi_R10_1.JPG *License*: unknown *Contributors*: Fabrice Pluchet

Image:Porsche 908 Langheck am 1981-08-15.jpg *Source*: http://en.wikipedia.org/w/index.php?title=File:Porsche_908_Langheck_am_1981-08-15.jpg *License*: unknown *Contributors*: User:Spurzem

Image:1929 Bentley front 34 left 2.jpg *Source*: http://en.wikipedia.org/w/index.php?title=File:1929_Bentley_front_34_left_2.jpg *License*: unknown *Contributors*: Chief tin cloud, Common Good, Erika39, Frank C. Müller, Herrick, Morio, Sfoskett

Image:Tom Kristensen 2008.JPG *Source*: http://en.wikipedia.org/w/index.php?title=File:Tom_Kristensen_2008.JPG *License*: unknown *Contributors*: User:Erika39

Image:Albert Guyot.jpg *Source*: http://en.wikipedia.org/w/index.php?title=File:Albert_Guyot.jpg *License*: unknown *Contributors*: Unknown (Bain News Service, publisher)

Image:Marcel Renault 1903.jpg *Source*: http://en.wikipedia.org/w/index.php?title=File:Marcel_Renault_1903.jpg *License*: unknown *Contributors*: Bukk, Fabriced28, Frank C. Müller, Infrogmation, Mattes, Morio, R17tsi, TFCforever, 2 anonymous edits

Image:Boillot-ACF-GP1912.jpg *Source*: http://en.wikipedia.org/w/index.php?title=File:Boillot-ACF-GP1912.jpg *License*: unknown *Contributors*: Gallant g, Markus3, Morio, Olivier2, SM

File:Bundesarchiv Bild 102-01319, Italien, Monza, Autorennen.jpg *Source*: http://en.wikipedia.org/w/index.php?title=File:Bundesarchiv_Bild_102-01319,_Italien,_Monza,_Autorennen.jpg *License*: unknown *Contributors*: Albertomos, G.dallorto, Gallant g, Ligabo, Luigi Chiesa, Morio, Pava, Pibwl, Typ932

Image:BenzTeardrop1923.jpg *Source*: http://en.wikipedia.org/w/index.php?title=File:BenzTeardrop1923.jpg *License*: unknown *Contributors*: Original uploader was JillandJack at en.wikipedia

File:US flag 48 stars.svg *Source*: http://en.wikipedia.org/w/index.php?title=File:US_flag_48_stars.svg *License*: unknown *Contributors*: AnonMoos, Clindberg, Dual Freq, Flargman4, Homo lupus, Jacobolus, MuXXo, Rocket000, Tkgd2007, Zscout370, 6 anonymous edits

File:Flag of Belgium (civil).svg *Source*: http://en.wikipedia.org/w/index.php?title=File:Flag_of_Belgium_(civil).svg *License*: unknown *Contributors*: Bean49, David Descamps, Dbenbenn, Denelson83, Evanc0912, Fry1989, Gabriel trzy, Howcome, IvanOS, Ms2ger, Nightstallion, Oreo Priest, Rocket000, Rodejong, Sir Iain, ThomasPusch, Warddr, Zscout370, 4 anonymous edits

File:Flag of Spain (1785-1873 and 1875-1931).svg *Source*: http://en.wikipedia.org/w/index.php?title=File:Flag_of_Spain_(1785-1873_and_1875-1931).svg *License*: unknown *Contributors*: User:HansenBCN, User:Ignaciogavira, User:SanchoPanzaXXI

File:Flag of the NSDAP (1920–1945).svg *Source*: http://en.wikipedia.org/w/index.php?title=File:Flag_of_the_NSDAP_(1920–1945).svg *License*: unknown *Contributors*: -

File:Flag of Germany (3-2 aspect ratio).svg *Source*: http://en.wikipedia.org/w/index.php?title=File:Flag_of_Germany_(3-2_aspect_ratio).svg *License*: unknown *Contributors*: User:Mmxx

File:National flag of Germany 1933-1935.svg *Source*: http://en.wikipedia.org/w/index.php?title=File:National_flag_of_Germany_1933-1935.svg *License*: unknown *Contributors*: User:SeNeKa

File:Flag of Switzerland.svg *Source*: http://en.wikipedia.org/w/index.php?title=File:Flag_of_Switzerland.svg *License*: unknown *Contributors*: User:-xfi-, User:Marc Mongenet, User:Zscout370

File:Flag of Spain 1931 1939.svg *Source*: http://en.wikipedia.org/w/index.php?title=File:Flag_of_Spain_1931_1939.svg *License*: unknown *Contributors*: User:SanchoPanzaXXI

File:Flag of Italy.svg *Source*: http://en.wikipedia.org/w/index.php?title=File:Flag_of_Italy.svg *License*: unknown *Contributors*: Anomie

File:Flag of the German Empire.svg *Source*: http://en.wikipedia.org/w/index.php?title=File:Flag_of_the_German_Empire.svg *License*: unknown *Contributors*: User:B1mbo, User:Madden

File:Flag of Czechoslovakia.svg *Source*: http://en.wikipedia.org/w/index.php?title=File:Flag_of_Czechoslovakia.svg *License*: unknown *Contributors*: (of code)

File:Flag of Hungary.svg *Source*: http://en.wikipedia.org/w/index.php?title=File:Flag_of_Hungary.svg *License*: unknown *Contributors*: User:SKopp

File:Flag of Chile.svg *Source*: http://en.wikipedia.org/w/index.php?title=File:Flag_of_Chile.svg *License*: unknown *Contributors*: User:SKopp

File:WWI-re.png *Source*: http://en.wikipedia.org/w/index.php?title=File:WWI-re.png *License*: unknown *Contributors*: Thomashwang

File:HMS Dreadnought 1906 H61017.jpg *Source*: http://en.wikipedia.org/w/index.php?title=File:HMS_Dreadnought_1906_H61017.jpg *License*: unknown *Contributors*: U.S. Navy

File:Gavrilo Princip captured in Sarajevo 1914.jpg *Source*: http://en.wikipedia.org/w/index.php?title=File:Gavrilo_Princip_captured_in_Sarajevo_1914.jpg *License*: unknown *Contributors*: unknown-anonymous

File:Austria Hungary ethnic.svg *Source*: http://en.wikipedia.org/w/index.php?title=File:Austria_Hungary_ethnic.svg *License*: unknown *Contributors*: Andrei nacu

File:Lettow's surrender.jpg *Source*: http://en.wikipedia.org/w/index.php?title=File:Lettow's_surrender.jpg *License*: unknown *Contributors*: anonymous African artist

File:TelegramWW1.jpg *Source*: http://en.wikipedia.org/w/index.php?title=File:TelegramWW1.jpg *License*: unknown *Contributors*: Austro-Hungarian government

File:Serbian Artillery WW1.jpg *Source*: http://en.wikipedia.org/w/index.php?title=File:Serbian_Artillery_WW1.jpg *License*: unknown *Contributors*: Lord Eru, 1 anonymous edits

File:German soldiers in a railroad car on the way to the front during early World War I, taken in 1914. Taken from greatwar.nl site.jpg *Source*: http://en.wikipedia.org/w/index.php?title=File:German_soldiers_in_a_railroad_car_on_the_way_to_the_front_during_early_World_War_I,_taken_in_1914._Taken_from_greatwar.nl_site.jpg *License*: unknown *Contributors*: Unknown German war photographer. Original uploader was Sus scrofa at en.wikipedia

File:WinstonChurchill1916Army.gif *Source*: http://en.wikipedia.org/w/index.php?title=File:WinstonChurchill1916Army.gif *License*: unknown *Contributors*: 1970gemini, Vanished user 001

File:Sunlight Soap WW 1 Ad.jpg *Source*: http://en.wikipedia.org/w/index.php?title=File:Sunlight_Soap_WW_1_Ad.jpg *License*: unknown *Contributors*: unknown

Image:Melbourne recruiting WWI.jpg *Source*: http://en.wikipedia.org/w/index.php?title=File:Melbourne_recruiting_WWI.jpg *License*: unknown *Contributors*: AustralianRupert, Mattes, Rcbutcher, Skeezix1000

File:Royal Irish Rifles ration party Somme July 1916.jpg *Source*: http://en.wikipedia.org/w/index.php?title=File:Royal_Irish_Rifles_ration_party_Somme_July_1916.jpg *License*: unknown *Contributors*: Royal Engineers No 1 Printing Company.

File:Canadian tank and soldiers Vimy 1917.jpg *Source*: http://en.wikipedia.org/w/index.php?title=File:Canadian_tank_and_soldiers_Vimy_1917.jpg *License*: unknown *Contributors*: Canada. Dept. of National Defence

File:GermanInfantry1914.jpg *Source*: http://en.wikipedia.org/w/index.php?title=File:GermanInfantry1914.jpg *License*: unknown *Contributors*: Madmax32, Patrolmanno9, Rcbutcher, 1 anonymous edits

File:BC RGA Leaders In England.jpg *Source*: http://en.wikipedia.org/w/index.php?title=File:BC_RGA_Leaders_In_England.jpg *License*: unknown *Contributors*: Magog the Ogre, Sreejithk2000

File:Grand Fleet sails.jpg *Source*: http://en.wikipedia.org/w/index.php?title=File:Grand_Fleet_sails.jpg *License*: unknown *Contributors*: Haza-w, Lothar von Richthofen, SkonesMickLoud, Wildsurmise, 6 anonymous edits

File:Hochseeflotte 2.jpg *Source*: http://en.wikipedia.org/w/index.php?title=File:Hochseeflotte_2.jpg *License*: unknown *Contributors*: Parsecboy, Sidonius, 1 anonymous edits

File:NationaalArchief uboat155London.jpg *Source*: http://en.wikipedia.org/w/index.php?title=File:NationaalArchief_uboat155London.jpg *License*: unknown *Contributors*: Docu, FlickreviewR, Ingolfson, Oneblackline, PeterJewell, Rcbutcher, Xaphire, 1 anonymous edits

File:Bulgaria southern front.jpg *Source*: http://en.wikipedia.org/w/index.php?title=File:Bulgaria_southern_front.jpg *License*: unknown *Contributors*: Unknown

File:Capture of Jerusalem 1917d.jpg *Source*: http://en.wikipedia.org/w/index.php?title=File:Capture_of_Jerusalem_1917d.jpg *License*: unknown *Contributors*: American Colony Photo Department (Jerusalem), photographer not named

File:Sarikam.jpg *Source*: http://en.wikipedia.org/w/index.php?title=File:Sarikam.jpg *License*: unknown *Contributors*: Charismaniac, Giorgiomonteforti, Hetoum I, Rcbutcher, Takabeg, WordLived

Image:Bundesarchiv Bild 146-1977-101-36, Deutsche Soldaten in Jerusalem.jpg *Source*: http://en.wikipedia.org/w/index.php?title=File:Bundesarchiv_Bild_146-1977-101-36,_Deutsche_Soldaten_in_Jerusalem.jpg *License*: unknown *Contributors*: Greenshed, Multichill, Netanel h, Rcbutcher, Túrelio, 6 anonymous edits

File:Austro-Hungarian mountain corps.jpg *Source*: http://en.wikipedia.org/w/index.php?title=File:Austro-Hungarian_mountain_corps.jpg *License*: unknown *Contributors*: Unknown Austro-Hungarian officer

File:Kämpfe auf dem Doberdo.JPG *Source*: http://en.wikipedia.org/w/index.php?title=File:Kämpfe_auf_dem_Doberdo.JPG *License*: unknown *Contributors*: R.A. Höger 1873-1930

Image:Marshal Joffre inspecting Romanian troops during WWI.jpg *Source*: http://en.wikipedia.org/w/index.php?title=File:Marshal_Joffre_inspecting_Romanian_troops_during_WWI.jpg *License*: unknown *Contributors*: Gheorghe Ionescu/Constantin Ivanovici/Tudor Posmantir/Eftimie Vasilescu/Nicolae Barbelian

File:Russian Troops NGM-v31-p379.jpg *Source*: http://en.wikipedia.org/w/index.php?title=File:Russian_Troops_NGM-v31-p379.jpg *License*: unknown *Contributors*: Conscious, EugeneZelenko, Jarekt, Martin H., Moverton, Petri Krohn, Rcbutcher, SoLando, 1 anonymous edits

File:Lenin.WWI.JPG *Source*: http://en.wikipedia.org/w/index.php?title=File:Lenin.WWI.JPG *License*: unknown *Contributors*: Davepape, Ertly, Jaro.p, Kaganer, Kl833x9, Kneiphof, Maksim, Svencb, 3 anonymous edits

File:Russian cavalry.jpg *Source*: http://en.wikipedia.org/w/index.php?title=File:Russian_cavalry.jpg *License*: unknown *Contributors*: Agence Rol

File:Brest-litovsk treaty.jpg *Source*: http://en.wikipedia.org/w/index.php?title=File:Brest-litovsk_treaty.jpg *License*: unknown *Contributors*: Magog the Ogre, Roxanna

File:River Crossing NGM-v31-p338.jpg *Source*: http://en.wikipedia.org/w/index.php?title=File:River_Crossing_NGM-v31-p338.jpg *License*: unknown *Contributors*: -

File:General gouraud french army world war i machinegun marne 1918.JPEG *Source*: http://en.wikipedia.org/w/index.php?title=File:General_gouraud_french_army_world_war_i_machinegun_marne_1918.JPEG *License*: unknown *Contributors*: US War Dept.

File:Bundesarchiv Bild 183-1983-0323-501, Kriegskinematograph im Schützengraben.jpg *Source*: http://en.wikipedia.org/w/index.php?title=File:Bundesarchiv_Bild_183-1983-0323-501,_Kriegskinematograph_im_Schützengraben.jpg *License*: unknown *Contributors*: Burts, Duch.seb, Rama, Rcbutcher

File:Guetteur au poste de l'écluse 26.jpg *Source*: http://en.wikipedia.org/w/index.php?title=File:Guetteur_au_poste_de_l'écluse_26.jpg *License*: unknown *Contributors*: Cohesion, Ejdzej, Florival fr, Gerbrant, Hohum, Itsmine, LeadSongDog, Petri Krohn, Rcbutcher, Soerfm, 2 anonymous edits

File:USA bryter de diplomatiska förbindelserna med Tyskland 3 februari 1917.jpg *Source*: http://en.wikipedia.org/w/index.php?title=File:USA_bryter_de_diplomatiska_förbindelserna_med_Tyskland_3_februari_1917.jpg *License*: unknown *Contributors*: Photograph by Harris & Ewing.

Image:Piave Front 1918.JPEG *Source*: http://en.wikipedia.org/w/index.php?title=File:Piave_Front_1918.JPEG *License*: unknown *Contributors*: Sgt. A. Marcioni

File:At close grips2.jpg *Source*: http://en.wikipedia.org/w/index.php?title=File:At_close_grips2.jpg *License*: unknown *Contributors*: H.D. Gridwood

File:British 55th Division gas casualties 10 April 1918.jpg *Source*: http://en.wikipedia.org/w/index.php?title=File:British_55th_Division_gas_casualties_10_April_1918.jpg *License*: unknown *Contributors*: 2nd Lt. T.L. Aitken

File:Aerial view of ruins of Vaux, France, 1918, ca. 03-1918 - ca. 11-1918 - NARA - 512862.tif *Source*: http://en.wikipedia.org/w/index.php?title=File:Aerial_view_of_ruins_of_Vaux,_France,_1918,_ca._03-1918_-_ca._11-1918_-_NARA_-_512862.tif *License*: unknown *Contributors*: Cobatfor, Dominic

File:American troops in Vladivostok 1918 HD-SN-99-02013.JPEG *Source*: http://en.wikipedia.org/w/index.php?title=File:American_troops_in_Vladivostok_1918_HD-SN-99-02013.JPEG *License*: unknown *Contributors*: Adam Zábranský, AnRo0002, Avron, BrokenSphere, Demidow, Dhatfield, Foreverprovence, KTo288, Mattes, 2 anonymous edits

File:World War I Observation Balloon HD-SN-99-02269.JPEG *Source*: http://en.wikipedia.org/w/index.php?title=File:World_War_I_Observation_Balloon_HD-SN-99-02269.JPEG *License*: unknown *Contributors*: Department of the Army. Office of the Chief Signal Officer - Signal Corps Photographs of American Military Activity.

File:US 64th regiment celebrate the Armistice.jpg *Source*: http://en.wikipedia.org/w/index.php?title=File:US_64th_regiment_celebrate_the_Armistice.jpg *License*: unknown *Contributors*: U.S. Army

File:Waffenstillstand gr.jpg *Source*: http://en.wikipedia.org/w/index.php?title=File:Waffenstillstand_gr.jpg *License*: unknown *Contributors*: 1970gemini, Duch.seb, Falkmart, Ferengi, Haukurth, Jef-Infojef, Johsan, Kresspahl, Labattblueboy, Maxim, Mutter Erde, Pieter Kuiper, Rcbutcher, TCY, Túrelio, Xenophon, 16 anonymous edits

File:Armisticetrain.jpg *Source*: http://en.wikipedia.org/w/index.php?title=File:Armisticetrain.jpg *License*: unknown *Contributors*: Andros64, Carl Logan, FranksValli, Quoique, Rcbutcher, TCY, 2 anonymous edits

File:Armoured car.jpg *Source*: http://en.wikipedia.org/w/index.php?title=File:Armoured_car.jpg *License*: unknown *Contributors*: Hohum, Quibik, Robin63

File:Sopwith F-1 Camel.jpg *Source*: http://en.wikipedia.org/w/index.php?title=File:Sopwith_F-1_Camel.jpg *License*: unknown *Contributors*: David Legrand, Dryke, Makthorpe, PMG, Stahlkocher, Wikipeder

File:Nieuport w Le Prieur rockets.jpg *Source*: http://en.wikipedia.org/w/index.php?title=File:Nieuport_w_Le_Prieur_rockets.jpg *License*: unknown *Contributors*: USAF

File:Battle of Messines - destroyed German trench.jpg *Source*: http://en.wikipedia.org/w/index.php?title=File:Battle_of_Messines_-_destroyed_German_trench.jpg *License*: unknown *Contributors*: Labattblueboy, LimoWreck, Rcbutcher, Soerfm

File:Vickers IWW.jpg *Source*: http://en.wikipedia.org/w/index.php?title=File:Vickers_IWW.jpg *License*: unknown *Contributors*: Breeze, Davepape, Denniss, Editor at Large, Jkelly, Krinkle, Kudlatek76, Nemo5576, Rcbutcher, Shimgray, Shizhao, SoLando, Thuresson, 15 anonymous edits

File:Hromadná poprava srbského obyvatelstva.jpg *Source*: http://en.wikipedia.org/w/index.php?title=File:Hromadná_poprava_srbského_obyvatelstva.jpg *License*: unknown *Contributors*: Boris Dimitrov, John commons, MarcusBritish, Miaow Miaow, 4 anonymous edits

File:Геноцид армян - 1915 год.jpg *Source*: http://en.wikipedia.org/w/index.php?title=File:Геноцид_армян_-_1915_год.jpg *License*: unknown *Contributors*: dabartinis

File:BVRC-Great-War-Contingent 1914.jpg *Source*: http://en.wikipedia.org/w/index.php?title=File:BVRC-Great-War-Contingent_1914.jpg *License*: unknown *Contributors*: British Army

File:Indian army soldier after siege of Kut q79446.jpg *Source*: http://en.wikipedia.org/w/index.php?title=File:Indian_army_soldier_after_siege_of_Kut_q79446.jpg *License*: unknown *Contributors*: Coyau, Flamurai, Fæ, Man vyi, Manxruler, Martin H., PKM, Timeshifter, 1 anonymous edits

File:German prisoners in a French prison camp. French Pictorial Service., 1917 - 1919 - NARA - 533724.tif *Source*: http://en.wikipedia.org/w/index.php?title=File:German_prisoners_in_a_French_prison_camp._French_Pictorial_Service.,_1917_-_1919_-_NARA_-_533724.tif *License*: unknown *Contributors*: Cobatfor, Dominic

File:Old england first.jpg *Source*: http://en.wikipedia.org/w/index.php?title=File:Old_england_first.jpg *License*: unknown *Contributors*: . Uploaded in Russian Wikipedia by Vizu

Image:Horace Smith-Dorrien LOC.jpg *Source*: http://en.wikipedia.org/w/index.php?title=File:Horace_Smith-Dorrien_LOC.jpg *License*: unknown *Contributors*: Possibly "Bain News Service, publisher."

File:The Deserter.jpg *Source*: http://en.wikipedia.org/w/index.php?title=File:The_Deserter.jpg *License*: unknown *Contributors*: Nirvana2013, Tm

File:1917 - Execution à Verdun lors des mutineries.jpg *Source*: http://en.wikipedia.org/w/index.php?title=File:1917_-_Execution_à_Verdun_lors_des_mutineries.jpg *License*: unknown *Contributors*: AndreasPraefcke, Emijrp, Juiced lemon, KoS, Quibik, Sherbrooke, Siebrand, Svencb, Timeshifter, 1 anonymous edits

File:Poprava vůdců rumburské vzpoury 1918.jpg *Source*: http://en.wikipedia.org/w/index.php?title=File:Poprava_vůdců_rumburské_vzpoury_1918.jpg *License*: unknown *Contributors*: Daniel Baránek, John commons, Miaow Miaow

File:Bundesarchiv Bild 183-R72520, Kiel, Novemberrevolution, Matrosenaufstand.jpg *Source*: http://en.wikipedia.org/w/index.php?title=File:Bundesarchiv_Bild_183-R72520,_Kiel,_Novemberrevolution,_Matrosenaufstand.jpg *License*: unknown *Contributors*: Duch.seb

File:0 Verdun - Cimetière de Douaumont (1).jpg *Source*: http://en.wikipedia.org/w/index.php?title=File:0_Verdun_-_Cimetière_de_Douaumont_(1).jpg *License*: unknown *Contributors*: User:Jean-Pol GRANDMONT

File:CampFunstonKS-InfluenzaHospital.jpg *Source*: http://en.wikipedia.org/w/index.php?title=File:CampFunstonKS-InfluenzaHospital.jpg *License*: unknown *Contributors*: U.S. Army photographer

File:Smyrna-massacre-refugees port-1922.jpg *Source*: http://en.wikipedia.org/w/index.php?title=File:Smyrna-massacre-refugees_port-1922.jpg *License*: unknown *Contributors*: JeromeKerviel, Sfrandzi

File:Beaumont hamel newfoundland memorial.jpg *Source*: http://en.wikipedia.org/w/index.php?title=File:Beaumont_hamel_newfoundland_memorial.jpg *License*: unknown *Contributors*: Citypeek, Kilom691, Père Igor, Rcbutcher, Skeezix1000

File:John McCrae in uniform circa 1914.jpg *Source*: http://en.wikipedia.org/w/index.php?title=File:John_McCrae_in_uniform_circa_1914.jpg *License*: unknown *Contributors*: William Notman and Son

File:Pagny le Chateau monument morts 002b.jpg *Source*: http://en.wikipedia.org/w/index.php?title=File:Pagny_le_Chateau_monument_morts_002b.jpg *License*: unknown *Contributors*: User:Grondin

Image:Siegfried Sassoon by George Charles Beresford (1915).jpg *Source*: http://en.wikipedia.org/w/index.php?title=File:Siegfried_Sassoon_by_George_Charles_Beresford_(1915).jpg *License*: unknown *Contributors*: MGA73, Materialscientist, Rotational, Rrburke

File:Cover-of-book-for-WWI-veterans-by-William-Brown-Meloney-born-1878.jpg *Source*: http://en.wikipedia.org/w/index.php?title=File:Cover-of-book-for-WWI-veterans-by-William-Brown-Meloney-born-1878.jpg *License*: unknown *Contributors*: GeorgeLouis

File:The announcing of the armistice on November 11, 1918, was the occasion for a monster celebration in Philadelphia... - NARA - 533478.tif *Source*: http://en.wikipedia.org/w/index.php?title=File:The_announcing_of_the_armistice_on_November_11,_1918,_was_the_occasion_for_a_monster_celebration_in_Philadelphia..._-_NARA_-_533478.tif *License*: unknown *Contributors*: Dominic, 1 anonymous edits

File:Map Europe 1923-en.svg *Source*: http://en.wikipedia.org/w/index.php?title=File:Map_Europe_1923-en.svg *License*: unknown *Contributors*: User:Fluteflute, User:Historicair

File:Bundesarchiv Bild 102-00104, Inflation, Tapezieren mit Geldscheinen.jpg *Source*: http://en.wikipedia.org/w/index.php?title=File:Bundesarchiv_Bild_102-00104,_Inflation,_Tapezieren_mit_Geldscheinen.jpg *License*: unknown *Contributors*: Pahl, Georg

File:The Girl Behind the Gun 1915.jpg *Source*: http://en.wikipedia.org/w/index.php?title=File:The_Girl_Behind_the_Gun_1915.jpg *License*: unknown *Contributors*: Beria, LGR109, Maximilian Schönherr

Image:Hydraylic disc brake diagram.jpg *Source*: http://en.wikipedia.org/w/index.php?title=File:Hydraylic_disc_brake_diagram.jpg *License*: unknown *Contributors*: Original uploader was KDS4444 at en.wikipedia

File:Loire Indre Tours1 tango7174.jpg *Source*: http://en.wikipedia.org/w/index.php?title=File:Loire_Indre_Tours1_tango7174.jpg *License*: unknown *Contributors*: User:Tango7174

file:France location map-Regions and departements.svg *Source*: http://en.wikipedia.org/w/index.php?title=File:France_location_map-Regions_and_departements.svg *License*: unknown *Contributors*: User:Sting

File:Red pog.svg *Source*: http://en.wikipedia.org/w/index.php?title=File:Red_pog.svg *License*: unknown *Contributors*: Anomie

File:Blason tours 37.svg *Source*: http://en.wikipedia.org/w/index.php?title=File:Blason_tours_37.svg *License*: unknown *Contributors*: User:Manassas

File:Tours Cathedral Saint-Gatian.jpg *Source*: http://en.wikipedia.org/w/index.php?title=File:Tours_Cathedral_Saint-Gatian.jpg *License*: unknown *Contributors*: User:Goldmund100

File:Hotel de Ville Tours.jpg *Source*: http://en.wikipedia.org/w/index.php?title=File:Hotel_de_Ville_Tours.jpg *License*: unknown *Contributors*: User:Willtron

File:Tours Cathedral from Rue Lavoisier.jpg *Source*: http://en.wikipedia.org/w/index.php?title=File:Tours_Cathedral_from_Rue_Lavoisier.jpg *License*: unknown *Contributors*: User:Ozeye

File:Loire Indre Tours2 tango7174.jpg *Source*: http://en.wikipedia.org/w/index.php?title=File:Loire_Indre_Tours2_tango7174.jpg *License*: unknown *Contributors*: User:Tango7174

File:Tram Tours Maquette-exterieur1.JPG *Source*: http://en.wikipedia.org/w/index.php?title=File:Tram_Tours_Maquette-exterieur1.JPG *License*: unknown *Contributors*: User:Tatmouss

File:LeoDupont.jpg *Source*: http://en.wikipedia.org/w/index.php?title=File:LeoDupont.jpg *License*: unknown *Contributors*: History2007

File:Flag of Germany.svg *Source*: http://en.wikipedia.org/w/index.php?title=File:Flag_of_Germany.svg *License*: unknown *Contributors*: Anomie

File:Flag of Spain.svg *Source*: http://en.wikipedia.org/w/index.php?title=File:Flag_of_Spain.svg *License*: unknown *Contributors*: Anomie

File:Flag of the People's Republic of China.svg *Source*: http://en.wikipedia.org/w/index.php?title=File:Flag_of_the_People's_Republic_of_China.svg *License*: unknown *Contributors*: User:Denelson83, User:SKopp, User:Shizhao, User:Zscout370

File:Flag of the United States.svg *Source*: http://en.wikipedia.org/w/index.php?title=File:Flag_of_the_United_States.svg *License*: unknown *Contributors*: Anomie

File:Flag of Canada.svg *Source*: http://en.wikipedia.org/w/index.php?title=File:Flag_of_Canada.svg *License*: unknown *Contributors*: Anomie

File:Flag of Japan.svg *Source*: http://en.wikipedia.org/w/index.php?title=File:Flag_of_Japan.svg *License*: unknown *Contributors*: Anomie

File:Flag of Romania.svg *Source*: http://en.wikipedia.org/w/index.php?title=File:Flag_of_Romania.svg *License*: unknown *Contributors*: User:AdiJapan

File:Public garden in Tours, France.jpg *Source*: http://en.wikipedia.org/w/index.php?title=File:Public_garden_in_Tours,_France.jpg *License*: unknown *Contributors*: AndreasPraefcke, Croquant, Erin Silversmith, Sir Gawain, TL

File:Giant Cypress tree in Tours, France.jpg *Source*: http://en.wikipedia.org/w/index.php?title=File:Giant_Cypress_tree_in_Tours,_France.jpg *License*: unknown *Contributors*: EmmanuelFrance, Erin Silversmith, MPF, Man vyi, Olybrius, TL, WayneRay, 1 anonymous edits

File:Hotel_gouin.JPG *Source*: http://en.wikipedia.org/w/index.php?title=File:Hotel_gouin.JPG *License*: unknown *Contributors*: BRUNNER Emmanuel, manu25

File:Tudor buildings in Tours, France.jpg *Source*: http://en.wikipedia.org/w/index.php?title=File:Tudor_buildings_in_Tours,_France.jpg *License*: unknown *Contributors*: AnRo0002, Balbo, Denniss, Erin Silversmith, Herbert Ortner, Kved, Mahlum, Makthorpe, Man vyi, Mifter, Olivier2, TL

File:Loire Indre Tours3 tango7174.jpg *Source*: http://en.wikipedia.org/w/index.php?title=File:Loire_Indre_Tours3_tango7174.jpg *License*: unknown *Contributors*: User:Tango7174

File:Rooftops of Tours, France.jpg *Source*: http://en.wikipedia.org/w/index.php?title=File:Rooftops_of_Tours,_France.jpg *License*: unknown *Contributors*: Denniss, Erin Silversmith, Makthorpe, Man vyi, Mifter, Olivier2, Ronaldino, TL, TomAlt

File:Pont Mirabeau Tours from the rivers north bank.jpg *Source*: http://en.wikipedia.org/w/index.php?title=File:Pont_Mirabeau_Tours_from_the_rivers_north_bank.jpg *License*: unknown *Contributors*: User:Ozeye

MIX
Papier aus verantwortungsvollen Quellen
Paper from responsible sources
FSC® C105338

Printed by Books on Demand GmbH, Norderstedt / Germany